Food Security and Poverty Reduction: Challenges and Opportunities for Farmers

Food Security and Poverty Reduction: Challenges and Opportunities for Farmers

Edited by Francis Hopkinson

SYRAWOOD
PUBLISHING HOUSE
New York

Published by Syrawood Publishing House,
750 Third Avenue, 9th Floor,
New York, NY 10017, USA
www.syrawoodpublishinghouse.com

Food Security and Poverty Reduction: Challenges and Opportunities for Farmers
Edited by Francis Hopkinson

International Standard Book Number: 978-1-64740-346-1 (Hardback)

Cataloging-in-publication Data

Food security and poverty reduction : challenges and opportunities for farmers / edited by Francis Hopkinson.
p. cm.
Includes bibliographical references and index.
ISBN 978-1-64740-346-1
1. Farms, Small. 2. Food security. 3. Food supply. 4. Agriculture. I. Hopkinson, Francis.
HD1476.A3 F66 2023
338.1--dc23

Table of Contents

PREFACE

I am honored to present to you this unique book which encompasses the most up-to-date data in the field. I was extremely pleased to get this opportunity of editing the work of experts from across the globe. I have also written papers in this field and researched the various aspects revolving around the progress of the discipline. I have tried to unify my knowledge along with that of stalwarts from every corner of the world, to produce a text which not only benefits the readers but also facilitates the growth of the field.

Agriculture refers to the science of cultivating the soil, growing crops and raising livestock. Carrying out agricultural activities in an efficient way can help a country meet or exceed its food security needs as well as in alleviating poverty. Farming can potentially play an important role in driving agricultural productivity. Poverty and hunger pose significant challenges in existing times. Food security refers to a situation that exists when all people have physical, social and economic access to sufficient, safe and nutritious food that meets their dietary needs for an active and healthy life. An increase in agriculture productivity not only increases food security, but also opens a plethora of employment opportunities outside of agriculture sector such as transport, agroprocessing, marketing, distribution and retail. The research works included in this book on food security and poverty reduction are of utmost significance and bound to provide incredible insights to readers. It is meant for students who are looking for an elaborate reference text on this topic.

Finally, I would like to thank all the contributing authors for their valuable time and contributions. This book would not have been possible without their efforts. I would also like to thank my friends and family for their constant support.

Editor

1

Inorganic Fertiliser Use Among Smallholder Farmers in Sub-Saharan Africa: Implications for Input Subsidy Policies

Jacob Ricker-Gilbert

Abstract In recent years, use of inorganic fertiliser among smallholder farm households in Sub-Saharan Africa (SSA) has increased, in large part due to the scale-up of input subsidy programmes (ISPs). However, fertiliser use efficiency for maize remains low, so the benefits of ISPs are often less than their costs. In order to make ISPs more cost-effective, sustainable and beneficial to smallholders, governments who implement ISPs should move towards implementing self-targeting mechanisms where more productive farmers opt into participation and relatively less productive farmers opt out. Such mechanisms include (i) increasing the amount of money that beneficiaries are required to contribute to acquire subsidised inputs and (ii) making receipt of subsidised inputs conditional on a household's willingness to implement soil fertility management practices that can increase the amount of maize produced per kilogram of fertiliser in future. Limited resource farmers who cannot provide complementary inputs to subsidised fertiliser would likely be better served by a cash transfer programme rather than an input subsidy.

1 Smallholder Access to Inorganic Fertiliser

Nutrients such as nitrogen, phosphorous and potassium are key inputs into the production of cereals, including maize which is the most widely produced and consumed cereal in SSA. The most effective mechanism to deliver these nutrients to crops is through the application of inorganic fertiliser (Vanlauwe et al. 2011). However, there is a common perception that farmers in SSA use significantly less fertiliser than is economically optimal. This belief has spurred significant research into the constraints

This chapter is derived from two original review reports written by the author and commissioned by the European Commission in 2015, entitled 'Smallholder Farmers' Access to Financial Instruments in Sub-Saharan Africa: A Review of the Evidence on Irrigation, Seeds, and Fertilizer', and by the World Bank in 2016, entitled 'Review of Malawi's Farm Input Subsidy Program in 2016 and Direction for Re-design'.

J. Ricker-Gilbert (✉)
Department of Agricultural Economics, Purdue University, West Lafayette, IN, USA
e-mail: jrickerg@purdue.edu

that inhibit and limit smallholders from using fertiliser. Reasons commonly given include supply-side problems such as poor infrastructure, late delivery of fertiliser, few input suppliers and inappropriate fertiliser blending and application rate recommendations that do not conform to local soil qualities (Gregory and Bumb 2006). Demand-side constraints include lack of credit at planting as a major inhibitor to using fertiliser, as identified by a number of studies (Coady 1995; Dorward et al. 2004; Duflo et al. 2011). Other studies point to unfavourable fertiliser/maize price ratios (Croppenstedt et al. 2003; Duflo et al. 2008) and poor soil quality leading to low maize to fertiliser response rates (Marenya and Barrett 2009), as reasons for low uptake of fertiliser.

With these considerations in mind, this chapter presents what is currently known about access to and use of inorganic fertiliser among smallholder farm households in SSA. I first discuss the common perceptions of low level of inorganic fertiliser use in SSA, and how recent data suggest that this may no longer be the case in many parts of the region. A substantial portion of the report is then devoted to discussing the challenges associated with input subsidy programmes (ISPs) that are currently being promoted by numerous governments in SSA to encourage inorganic fertiliser use among smallholders.[1] Evidence indicates that by lowering the price of fertiliser for smallholders, ISPs have contributed to increased fertiliser application per hectare among smallholders. However, low maize to nitrogen response rates are a major challenge for the cost-effectiveness of these programmes. I discuss the problem of low maize to fertiliser response rates and factors that explain why this occurs. I conclude with policy recommendations for making ISPs more cost-effective, sustainable and beneficial to smallholder farm households in SSA.

1.1 Background: Common Perception of Low Inorganic Fertiliser Use in SSA

There remains a common perception that inorganic fertiliser use among smallholder farm households in SSA is extremely low. Aggregate, national-level data from FAOSTAT suggest that on average, farmers across the region use only 13 kg of fertiliser nutrients per hectare of arable land, which is far below the developing-country average of 94 kg/ha (Minot and Benson 2009). Low fertiliser use, low yields, persistent poverty, along with several food price spikes over the past ten years have increased awareness of the need to increase smallholder staple crop production in SSA.

As a result, numerous African policymakers met in Abuja, Nigeria in 2006 at the African Fertilizer Summit, where they vowed to help smallholder farmers to access inorganic fertiliser as the primary mechanism for increasing agricultural productivity. The main policy mechanism advocated was through targeted ISPs. In targeted

[1] According to Jayne and Rashid (2013), seven countries in SSA spent the equivalent of USD 1.05 billion in 2011 subsidising inputs. This is equivalent to 28.6% of public spending on agriculture (Jayne and Rashid 2013).

subsidy programmes, a sub-sample of farm households who meet certain criteria is able to acquire a limited quantity of inorganic fertiliser at a price below the market price (subsidy). These targeted programmes in theory are supposed to overcome the problems with universal fertiliser subsidies that were common across SSA in the 1970s and 1980s. Under universal subsidy programmes, the government controls the price of fertiliser and makes it available to all farmers at a below-market rate. Evidence suggests that most of the benefits from universal subsidies go to wealthier households who are better able to access the fertiliser and to input suppliers who do not fully pass the cost savings back to farmers (Brooks et al. 2008). In addition, due to their high costs, universal subsidy programmes became financially untenable for many countries and were phased out in the late 1980s and 1990s under structural adjustment.

Recent evidence from nationally representative household-level panel survey data from six countries in SSA, collected by the World Bank in the years following input subsidy scale-up in many countries, indicates that inorganic fertiliser use may not be as low as commonly perceived. Table 1 (Sheahan and Barrett 2014) shows that the average rate of inorganic fertiliser nutrients is 26 kg/ha, double the 13 kg/ha from FAOSTAT data. In addition, several countries have nutrient use rates for inorganic fertiliser that are significantly higher than 26 kg/ha. It is also not surprising that the countries with the highest rates of inorganic fertiliser use—Malawi, Nigeria and Ethiopia—have all funded large fertiliser subsidy programmes in recent years. Malawi and Nigeria administer a targeted ISP, while Ethiopia uses a universal subsidy programme where the government imports fertiliser and distributes it at below-market price to farmers across the country, through its network of cooperative unions.

Higher fertiliser use than commonly perceived is also seen in Malawi (Ricker-Gilbert and Jayne 2017). This study follows 462 smallholder Malawian farm households over eight years. Fertiliser use patterns for these households are presented in Fig. 1. The figure demonstrates that the ISP in Malawi has contributed to raising fertiliser use since it was first scaled up during the 2005/06 season. Average fertiliser use

Table 1 Average household-level organic and inorganic fertiliser use trends

Country	% of cultivating households using inorganic fertiliser	Use (kg/ha) of inorganic fertiliser across all households (includes zeros)	
		Mean total	Mean nutrients
Ethiopia	55.5	45.0	25.2
Malawi	77.3	146.0	56.3
Niger	17.0	4.5	1.7
Nigeria	41.4	128.2	64.3
Tanzania	16.9	16.2	7.7
Uganda	3.2	1.2	0.7
Average	35.2	56.9	26.0

Source Sheahan and Barrett (2014)

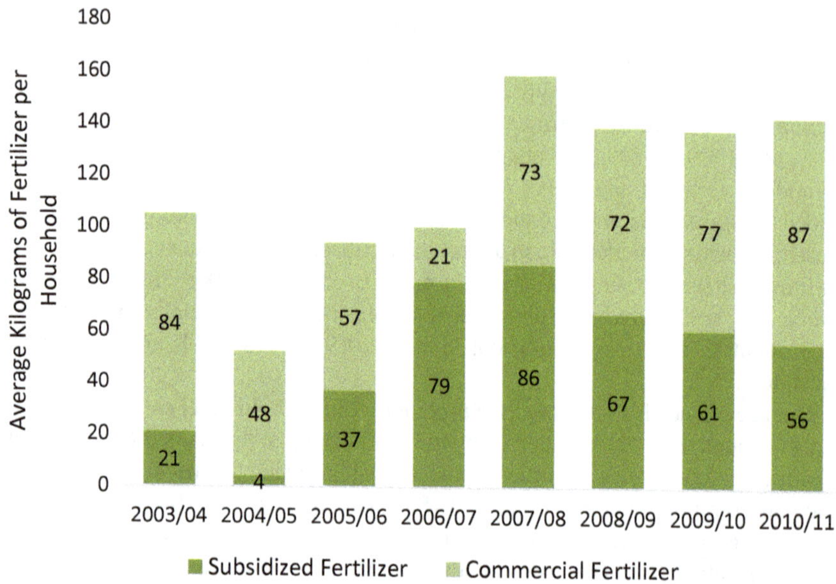

Fig. 1 Average fertiliser use (kg) by households in Malawi, by year and by source. $N = 462$ households in each year. *Source* Ricker-Gilbert and Jayne (2017)

per household stood at slightly more than 100 kg in 2003/04 but increased to nearly 150 kg in 2010/11. The figure also shows that commercial fertiliser use declined during the initial years when the subsidy was scaled up but has since rebounded to its pre-subsidy level.

1.2 Challenge of Low Maize to Fertiliser Response Rates

Although the evidence suggests that input subsidies have contributed to increasing fertiliser use among smallholders in SSA in recent years, fertiliser acquisition and use are just one component for raising yields and productivity. The first major challenge facing ISPs is making sure that recipients are using fertiliser efficiently so that the marginal benefits of using fertiliser are greater than the marginal costs.

The marginal product of fertiliser (kilograms of maize produced per kilogram of nitrogen) is a key factor determining whether or not the benefit/cost ratios for ISPs are greater than 1 and thus break even, or do better than that. Jayne and Rashid (2013) review the literature on ISPs in SSA and compile a table of studies across the region that estimate the marginal product of maize to nitrogen and benefit/cost ratios. Results from these studies are presented in Table 2. The main conclusion that can be drawn from this table is that the marginal product of fertiliser is quite low, and thus, the benefit/cost ratios hover around 1 or are below 1 in many contexts.

Table 2 Recent estimates of maize response to nitrogen applications in SSA

Study	Country	Agronomic response rate (kg maize per kg nitrogen)	Benefit/cost ratio
Minten et al. (2013)	Ethiopia	10–14	1.0–1.4
Sheahan et al. (2013)	Kenya	14–21	1.3–3.7
Marenya and Barrett (2009)	Kenya	17.6	1.76
Matsumoto and Yamano (2011)	Uganda	8.0	0.75–1.05
Burke (2012)	Zambia	9.6	0.3–1.2
Ricker-Gilbert and Jayne (2012)	Malawi	8.1	0.6–1.6
Pan and Christiaensen (2012)	Tanzania	11.7	–

Source Adapted from Jayne and Rashid (2013) and Burke et al. (2015)

This consistent finding raises questions about whether or not subsidies for fertiliser can by themselves be a cost-effective strategy for raising smallholder agricultural productivity.

1.3 Reasons for Low Maize to Fertiliser Response Rates

The next logical question to ask is why are response rates to fertiliser so low and what can be done to improve them, so that inorganic fertiliser is more profitable for smallholder farmers in SSA to use? The first potential challenge is, as mentioned earlier, that many farmers in SSA are dependent on rainfed agriculture and lack access to water control through irrigation. Water control is crucial for plant growth and for the economic returns on using fertiliser. Dependence on rainfall raises the risk associated with purchasing inorganic fertiliser, as climate trends suggest that most of SSA has been receiving less and more sporadic rainfall, and will continue to do so in future under most climate change scenarios (Niang et al. 2014). Access to irrigation affords more reliable water control and more stable yield response to fertiliser, compared to rainfed cultivation. Unfortunately, only 4% of arable land is under irrigation in SSA, compared to 45% in South Asia (Jayne and Rashid 2013). This difference helps to explain why fertiliser application rates and maize to fertiliser response rates are much lower in SSA than in South Asia.

The second reason for low maize to fertiliser response rates in SSA is poor and degrading soil quality. Rapid population growth in many parts of SSA leads to smaller and smaller farms that continuously cultivate cereals year after year with little nutrient replenishment, leading to worsening soil quality which in turn leads to lower yields. Marenya and Barrett (2009) demonstrate that, in western Kenya, soil organic matter (SOM) is an important indicator of soil degradation that has a strong effect on maize to fertiliser response rates. The authors conclude that, given low levels of SOM, it is not profitable for many smallholders to purchase inorganic fertiliser.

Intercropping maize with legumes is one relatively low-cost way for soil fertility to be maintained or perhaps restored. Legumes have the ability to fix nitrogen at a higher rate than cereals do, so their presence in a cropping system can help build nitrogen and organic matter over time (Snapp 1998). Unfortunately, the percentage of maize fields intercropped with legume is not as high as it could be, and there is some evidence that the rate of intercropping may be declining over time. For example, Snapp et al. (2014) show that in Malawi, 50.1% of maize plots were intercropped with legumes in 2002/03; this percentage declined to 46.1% in 2006/07, 45.4% in 2008/09 and 37.9% in 2009/10—a worrying trend.

The third reason for low response rates is late delivery and application of fertiliser. Proper timing of fertiliser application is important to prevent nutrient loss, increase nutrient use efficiency and prevent damage to the environment through nutrient run-off (Jones and Jacobsen 2003; Snapp et al. 2014). Xu et al. (2009) find that timely application of fertiliser is one of the major factors that has a positive impact on maize response to fertiliser in Zambia. However, it is unfortunately not the case that farmers always acquire and apply fertiliser at the appropriate time. Snapp et al. (2014) find that more than half of all smallholders in Malawi apply their first dosage of fertiliser more than three weeks after planting, which is generally later than optimal for yield maximisation. There could be various explanations for late application of fertiliser, such as late delivery to fertiliser retailers, smallholders lacking sufficient labour to apply the fertiliser and not having the management ability and knowledge to appropriately apply fertiliser.

The fourth reason for low response rates is appropriate management and timely weeding. Weeding is essential to improve the ability of plants to access and use nitrogen and phosphorous effectively. Repeated weeding of maize during a growing season is essential to maximise yields, and farmers who weed their maize only once during the growing season can experience a 26–34% decline in yields due to the build-up of weeds (FAO 2000). Pests such as the parasitic weed *striga* are a major challenge for many smallholders in SSA; they can cause major yield losses if not removed through weeding or herbicide application. Snapp et al. (2014) find that in Malawi only 65–70% of plots are weeded twice as recommended, while 25–27% of maize plots are weeded only once or not at all. Furthermore, the authors find that 13.7–17.3% of households say that they have experienced yield reductions due to crop diseases or pests over the past two to three years.

2 Implications for Input Subsidy Policy

Dependence on rainfed agriculture, poor and worsening soil quality, late delivery and application of fertiliser and insufficient weeding all help to explain the low maize to fertiliser response rates observed in the studies presented in Table 2. Low response rates are a major challenge for input subsidies and undermine their cost-effectiveness and sustainability in the long run. The issues highlighted above demonstrate that inorganic fertiliser is just one input into the production of cereals, which also depends on land, seed, water, labour, soil fertility and management ability. Therefore, there is a need for countries in SSA to move from a development strategy where substantial shares of national agricultural budgets are devoted to subsidising nitrogen and phosphorous, to a more holistic agricultural development strategy that focuses on soil fertility as a complement to inorganic fertiliser.

Unfortunately, in the past, focusing on soil fertility has sometimes been viewed as 'low input' or 'alternative' agriculture. However, research in the agronomy and soil science literature increasingly indicates that holistic soil fertility management (SFM) will be required to enable smallholders to use inorganic fertiliser more intensively and profitably. In this light, SFM and inorganic fertiliser can be viewed as complements that are necessary for one another, rather than substitutes that should take the place of each other. Elements of a holistic strategy would include (i) developing improved seeds that have the characteristics that farmers desire—accomplishing this would require more support for national agricultural research systems and (ii) increasing funding and support for extension programmes to help limited resource farmers to improve maize to fertiliser response rates. This could occur through better training in weeding and improved fertiliser management, along with programmes to restore soil fertility (Snapp et al. 2014).

2.1 *Need to Clarify ISP Goals and Objectives*

As mentioned previously, the need for complementary investments to raise maize to fertiliser response rates is the first challenge to making ISPs cost-effective and sustainable. The second challenge is ambiguity and expectations concerning the goal of these programmes. Take, for example, the goals of Malawi's Farm Input Subsidy Program (FISP), which are to increase productivity and reduce poverty by targeting the 'productive poor', who are broadly defined as full-time smallholder farmers who can contribute to increasing national-level production, but cannot afford to purchase one or two 50 kg bags of fertiliser at commercial prices (Dorward et al. 2008). This definition can be compared to the official targeting criteria for beneficiary selection under FISP, as of 2007/2008: (i) households headed by a Malawian who owns and currently cultivates land, (ii) vulnerable households, including guardians of physically challenged persons, and households headed by females, orphans or children. There is clear inconsistency between targeting the 'productive poor' and targeting vulnerable households because vulnerable households often do not have the land, labour and skills necessary to use inputs effectively. This inconsistency complicates both the evaluation of how well ISPs target the intended beneficiaries, and how effectively the FISP meets its stated objectives of increasing maize productivity, promoting household food security and reducing poverty.

As mentioned, fertiliser and seed subsidies require complementary inputs such as land, labour and management practices, so it makes sense for their goals to focus on helping smallholders to boost food production. However, due in part to their high cost and substantial budget share, many people expect that ISPs should be able to both increase food production and reduce household vulnerability to poverty and hunger. There may be some overlap between households who can increase maize production through input subsidies and households who have their vulnerability reduced through input subsidies.

2.2 *Improving Beneficiary Targeting*

Two inter-related challenges with ISPs are (i) what are the characteristics of intended beneficiaries? and (ii) how can they be effectively targeted? The following subsection identifies three different potential targeting methods for reaching intended ISP beneficiaries.

2.2.1 Community-Based Targeting

At the local level, many countries including Malawi, Zambia, Nigeria and Kenya rely on a decentralised targeting system where local chiefs determine who should receive coupons for subsidised seed and fertiliser. Community-based targeting programmes have the benefit of using local knowledge to identify beneficiaries at a relatively low cost to the government. However, community-based targeting schemes are more likely to suffer from elite capture, where those with social connections and resources obtain a disproportionate share of the benefits (Pan and Christiaensen 2012). The majority of evidence from Malawi's ISP indicates that over the programme's duration, Malawi's rural poor have not been specifically targeted to receive subsidised fertiliser and seed (Ricker-Gilbert et al. 2011; Holden and Lunduka 2012; Chibwana et al. 2011; Kilic et al. 2015). Kilic et al. recently found that on average relatively well-off households, who are connected to community leadership and reside in agro-ecologically favourable locations, are more likely to be ISP beneficiaries. However, Fisher and Kandiwa (2014) find evidence that, in recent years, female-headed households are significantly more likely to be targeted by the ISP.

The difficulty in determining whether or not ISPs effectively target intended beneficiaries relates back to the problem of clarifying the programme's goals, which affects who the intended beneficiaries should be. The fact that relatively better-off households in areas with favourable agro-ecology are more likely to obtain subsidised fertiliser in Malawi, suggests that the community-based system may target inputs towards more productive households. A recent working paper by Basurto et al. (2015) finds support for the notion that chiefs take productive efficiency into account when identifying beneficiaries. However, they find that chiefs are more likely to offer the inputs to relatives rather than non-relatives, consistent with the idea of elite capture in community-based targeting programmes.

In order to deal with this ambiguity in community-based targeting, it is important that programme goals and targeting guidelines are clarified and communicated with communities so that they understand the following households should be the primary beneficiaries of ISP: (i) households who are credit constrained and may lack cash at planting to purchase inorganic fertiliser at commercial prices, (ii) households who have sufficient land and labour to make use of the inorganic fertiliser and seed. It should also be made clear that village leaders and relatives of chiefs should not receive any preferential treatment for receiving subsidised inputs. Clarifying and communicating goals to communities will likely not solve all of the targeting challenges, but it can remove confusion about who beneficiaries should be and help create accountability within the communities.

2.2.2 Proxy-Means Targeting Based on Observables

Proxy-means (PM)-based targeting systems, where households are selected to participate in the programme based on observable characteristics that proxy for household need, are one potential alternative to a decentralised, community-based targeting system. The advantage of a PM system is that, in theory, the criteria for beneficiary selection are clear and observable to all parties involved in the programme, thus providing transparency and potentially reducing elite capture problems that exist in community-based systems. The challenge with PM targeting is that it can be difficult to find observable proxies that accurately represent beneficiary need. In addition, potential beneficiaries have the incentive to make observable proxies appear unobservable to government officials who distribute the benefits of the programme, once they understand that eligibility depends on observing or not observing these proxies.

To my knowledge, the only study to directly compare PM targeting with community-based targeting in Malawi is the working paper by Basurto et al. (2015). The authors find that FISP's community-based targeting system performs almost as well as an ideal PM system based on observable proxies for household consumption. The authors point out that the consumption information in their study is a more reliable predictor of need than household assets, which is what the government would have to rely on if the FISP moves to a large-scale PM-based system. This suggests that the current community-based targeting system may distribute FISP inputs more effectively than an asset-based PM system. In total, the results from Basurto, Dupas and Robinson indicate that the current community-based targeting system should not be replaced with a PM system. It is possible that an improved FISP targeting system could incorporate some combination of community-based and PM targeting schemes, but this would require further investigation and piloting.

2.2.3 Self-targeting Mechanisms

Given the challenges associated with identifying intended beneficiaries to be targeted by ISP using either community-based, PM or geographic targeting, serious consideration should be given to programme modifications that induce self-targeting of beneficiaries. With self-targeting, appropriate beneficiaries self-select into ISP participation, and inappropriate beneficiaries self-select out of participating. Since this report argues that ISP goals should be to increase maize yields and maize output, the programme should be modified to encourage self-selection among relatively more productive smallholders who can contribute to these goals.

The first and perhaps the most straightforward method for encouraging self-targeting for ISP participation is to increase the contribution that beneficiaries are required to make to redeem subsidised fertiliser. For example, in Malawi the government increased the required farmer contribution in 2015/16 to about 30% of the total cost of fertiliser, from the 5% that beneficiaries were paying in previous years. This price increase serves as a self-targeting mechanism since smallholders will only purchase fertiliser if they believe that the marginal benefits of using it will outweigh the marginal costs. Therefore, raising the effective price that beneficiaries pay for subsidised fertiliser will induce more productive smallholders to self-select into participation as they can obtain high enough returns to cover the increased costs of acquiring the input.[2] At the same time, less productive farmers who can cover their costs at a 95% subsidy rate will find it increasingly less profitable to participate as the rate of subsidy decreases, and their costs increase.

Furthermore, as the size of the required beneficiary contribution increases, the arbitrage opportunities for subsidy beneficiaries to re-sell subsidised fertilizer are reduced. Resale of subsidised fertiliser on a secondary market has been identified in several studies as a major problem with Malawi's subsidy programme in recent years (Holden and Lunduka 2010; Lunduka et al. 2013). Therefore, governments who currently subsidise fertiliser should consider increasing the required farmer contribution to the ISP by 10 percentage points per year, as a strategy for gradually graduating from the programme. Increasing the required farmer contribution to subsidised fertiliser would also relieve budgetary pressure from the programme. Savings from input subsidies could be channelled into programmes that can complement the inorganic seed and fertiliser use, by helping farmers to restore and improve soil fertility over time.

With this consideration in mind, the second self-targeting mechanism advocated in this report is for a conditional subsidy programme, where receiving subsidised fertiliser would be conditional on beneficiary smallholders making some form of investment in their long-term soil fertility. These investments could take the form of (i) planting nitrogen-fixing trees, such as *Faidherbia*; (ii) using contour ridging on their fields to prevent erosion; (iii) using organic manure to improve soil fertility

[2]This is consistent with a policy simulation that was recently conducted by the International Monetary Fund (IMF) (Nsengiyumva et al. 2015).

and soil structure and (iv) using pit planting to reduce nutrient run-off, and increase nutrient uptake by maize plants.

Adoption of these practices requires some labour and incurs monetary costs on the part of smallholders. A recent study in Malawi by Jack (2013) finds that farmers self-select into and out of a programme that encourages tree planting based on the individual smallholder's perception of future costs and benefits associated with planting and maintaining their trees. Therefore, if input subsidies change towards a conditional subsidy, then smallholders who are willing to incur the cost of this type of soil fertility investment can select into the programme in order to obtain subsidised fertiliser and seed. Conversely, smallholders who do not want to or are unable to make this investment will select out of the programme and choose to not receive the subsidised inputs.

A conditional subsidy would require some verification by extension officers or another third party, but tree planting is relatively easy to observe, so it may be a desirable intervention from a cost-effective oversight standpoint. In addition, the government will likely need to invest some resources in training farmers and providing extension advice on these proposed soil fertility improving practices. The additional costs of verification and farmer training could be paid in part by the savings incurred from increasing the required farmer contribution to the ISP.

2.3 ISPs Versus Other Social Protection Programmes

The following sub-section compares the ISP to other similar programmes that can be targeted or are targeted towards smallholders in SSA. This sub-section discusses what is known about cash transfers and flexible input voucher (FIV) programmes—two possible alternative mechanisms to ISPs.

2.3.1 Cash Transfer Programmes

Cash transfer programmes provide money directly to recipients. Unlike ISPs, cash transfers do not require beneficiaries to have complementary land and labour input to make use of them. In terms of effectiveness, targeted cash transfers share many of the problems that one finds with targeted ISPs. These problems include greater participation by individuals who have connections to local leaders and households 'gaming the system' to appear more needy than they actually are (Ellis and Maliro 2013). However, one would expect that the administrative burden of distributing a cash transfer would be lower than the burden for distributing subsidised inputs. Cash transfer programmes would likely be a more effective mechanism than ISPs for directly reaching limited resource beneficiaries to provide them with direct resources to reduce their vulnerability to hunger and poverty.

2.3.2 Flexible Input Voucher Programmes

Flexible input vouchers (FIVs) have a certain cash value associated with them and allow recipients to redeem them at an input supplier for whatever combination of inputs best suits their needs. In addition, the FIV is a potential way of supporting and strengthening the private network for input distribution, wholesale and retail in SSA countries. Unfortunately, there is little evidence on the household-level impacts of FIV programmes. To my knowledge, the only study to date to evaluate FIVs is based on a pilot programme in Zimbabwe. The study measures FIV impacts on input suppliers and how effectively the programme reaches recipient farmers (Mazvimavi et al. 2013). The authors find that FIVs help retailers to boost sales and revenue, and help to link farmers to input suppliers. They find that FIVs work better in areas with good infrastructure and good mobile phone reception. The authors also identify challenges associated with FIVs, such as getting retailers to stock the full complements of inputs that farmers may want. In addition, wholesalers face financial risks if not all agro-inputs are purchased.

There is currently insufficient evidence to recommend that countries in SSA move fully towards an FIV system. However, it may be worth piloting an FIV programme in a few districts to compare its impact against current systems. An evaluation of the e-voucher programme in Malawi by Tsoka et al. (2015) reveals that the electronic system removes the problem of fraudulent printing of counterfeit paper vouchers, lowers the administrative costs compared to paper vouchers and makes it easier and faster for the government to reimburse input suppliers, which helps suppliers to keep inputs in stock. Combining an FIV pilot with an e-voucher could help to ensure accountability and swift repayment for inputs from the government to private retailers. As mobile phone use increases among smallholders in SSA and coverage improves, e-voucher options will likely become more cost-effective and viable in future. Such a system could be a step towards graduating from input subsidies in Malawi.

3 Recommendations for Improving ISPs

There are a number of ways in which ISPs can be modified and improved to better target smallholders who can effectively use the inputs while relieving pressure on the government's budget and moving towards eventual graduation from the programme. Given the challenges associated with effectively targeting programme beneficiaries, steps should be taken to encourage self-targeting as much as possible. The following four recommendations are potential ways of improving ISPs throughout SSA.

3.1 *Governments Should Clearly Communicate that the Main Goal of ISPs Is to Increase Maize Yields and Maize Production*

Poverty reduction *should not* be thought of as a primary objective of the ISP. Clarifying programme goals and communicating them to local communities will help to remove ambiguity in the targeting guidelines and lower the expectations on ISPs to be successful in multiple dimensions. Clarifying programme objectives will also help communities to understand that the following households should be the primary beneficiaries of ISP: (i) households who are credit constrained and may lack cash at planting to purchase inorganic fertiliser and seed at commercial prices; and (ii) households who have sufficient land and labour to make use of the inorganic fertiliser and seed. Given evidence that ISPs have been prone to elite capture where better connected households are more likely to receive subsidised inputs, it should be clearly communicated to the village that kin of the village chief and other leaders in the community should not be any more likely to receive ISP benefits than other households. Clarification of this issue may help to create some self-enforcement within the community, but oversight by agricultural extension personnel may be needed for auditing purposes.

3.2 *Self-targeting Option No 1: Further Increase the Required Farmer Contribution to ISP Fertiliser*

Increasing the farmer contribution as a percentage of the total cost of the ISP is a strategy for gradually graduating from the programme. Increasing the required farmer contribution to receive subsidised inputs has three main programmatic benefits. (1) It lowers the cost of the ISP for the government and relieves pressure on the budget and balance of payments. In doing so, money that went to paying for ISP can be transferred to other complementary programmes. (2) It reduces the gains from reselling subsidised fertiliser on the secondary market. (3) It serves as a self-targeting mechanism. Since we assume that farmers will only purchase fertiliser if they believe that the marginal benefits of using it will outweigh the marginal costs, raising the effective price that farmers pay for subsidised fertiliser will induce more productive farmers, who can obtain a high enough return to cover the increased costs, to self-select into ISP participation. At the same time, less productive farmers who can cover their costs at a 95% subsidy rate will find it increasingly less profitable to participate as the rate of subsidy decreases.

3.3 *Self-targeting Option No 2: Pilot a Conditional Subsidy*

As discussed earlier, inorganic fertiliser is just one input into the production of maize. Fertile soils that maintain sufficient nitrogen, phosphorous and other nutrients are crucial to increasing nitrogen use efficiency (NUE) from inorganic fertiliser, thus increasing the cost-effectiveness of ISPs. Therefore, it would be advantageous to make receiving subsidised fertiliser and seed conditional on smallholders making some form of investment in their long-term soil fertility. A conditional subsidy would require some third-party verification, so planting trees would be an option that is relatively easy for extension agents or others to verify. Under a conditional subsidy, smallholders who are willing to make these investments in soil fertility can select into the programme in order to obtain subsidised fertiliser and seed, while smallholders who do not want to or are unable to make this investment will select out of the programme.

3.4 *Pilot Flexible Input Voucher (FIV) Programme*

FIVs have a certain cash value associated with them that allows recipient smallholders to redeem the value at an input supplier for whatever combination of inputs best suits their needs. FIVs allow the government to recognise that the 100 kg of fertiliser and 5–8 kg of maize seed distributed under FISP may not be the best for everyone. The FIV has two main programmatic benefits over the current FISP: (1) FIVs offer a level of empowerment where households get to choose what input or combination of inputs they want; (2) FIVs provide a potential mechanism for supporting and strengthening the private network for input distribution, wholesale and retail in the countries where they operate. By providing empowerment and flexibility for smallholder households and support to the private sector, a move to an FIV system could be a step towards graduating from input subsidies. Use of an e-voucher with FIV could help to ensure accountability and swift repayment for products from the government to private sector retailers.

4 Conclusions

The present chapter reviews the recent literature on smallholder access to and use of inorganic fertiliser in SSA and draws implications for input subsidy programmes (ISPs). It seems clear, as demonstrated in Sheahan and Barrett (2014) and Ricker-Gilbert and Jayne (2017), that ISPs have helped to increase inorganic fertiliser use among smallholders in SSA. However, there are two major challenges that threaten the cost-effectiveness and sustainability of these programmes. The first is that low response rates of maize to fertiliser and relatively high costs of implementing these programmes make it difficult for their marginal benefits to exceed their marginal costs. The second challenge is the need for ISPs to clarify their objectives and goals. Because fertiliser and seed subsidies require complementary inputs such as land and labour, these should be viewed primarily as productivity enhancing programmes. However, in part due to their high cost and the substantial share of national budgets allocated to them, they are often expected to also reduce poverty and vulnerability. This double burden puts tremendous pressure on ISPs.

These challenges make it necessary to find ways to make ISPs more effective and sustainable. In order to do so, this report makes the following set of recommendations. First, the goals of ISPs should be clarified, with a primary focus on enhancing productivity. Second, it is very difficult to determine, identify and target appropriate beneficiaries for the ISP using community-based targeting or proxy-means-based targeting schemes. Therefore, ISPs should be scaled down, and the government should move to implement self-targeting mechanisms where more productive farmers opt in and less productive farmers select opt of participating. One example of a self-targeting mechanism that is fairly easy to implement is to increase the required farmer contribution to the ISP. This would increase the marginal cost of acquiring subsidised fertiliser relative to the marginal benefit of using it. Therefore, as the cost to recipients increases, those who cannot make a profit at higher effective prices will likely self-select out. The second self-targeting mechanism that should be considered is making the receipt of subsidised inputs conditional on households who are willing to make longer-run investments in soil fertility. This means that if smallholders are willing to plant a tree or use organic manure or create contour ridging on their fields, they could be eligible to acquire subsidised fertiliser. Those who do not want to or are unable to make such investments will opt out of the programme. Requiring ISP recipients to make soil fertility investments will help to improve the soil fertility and the maize response to fertiliser over time on smallholder fields. Doing so will make ISPs more cost-effective and sustainable in the longer term.

While the ISP should focus on increasing maize productivity and production, cash transfer programmes to households should be scaled up, to provide income support that directly increases consumption for poorer households who cannot make effective use of subsidised inputs because they are land or labour constrained. Depending on availability of funds, a cash transfer could be scaled up as ISPs are scaled down.

References

Basurto, P., Dupas, P., & Robinson, J. (2015). *Decentralization and efficiency of subsidy targeting: Evidence from chiefs in rural Malawi* (Working Paper). Department of Economics, UC-Santa Cruz.

Brooks, J., Dyer, G., & Taylor, E. (2008). *Modelling agricultural trade and policy impacts in less developed countries* (OECD Food, Agriculture and Fisheries Working Papers 11).

Burke, W. (2012). Determinants of maize yield response to fertilizer application in Zambia: In Implications for strategies to promote smallholder productivity. (PhD Thesis), East Lansing, MI: Michigan State University.

Burke, W. J., Frossard, E., Kabwe, S., & Jayne, T. S. (2015). Soil management and crop response to fertilizer use in Sub-Saharan Africa: Methodological and policy issues: Examples from Zambia. In *Organized Symposia as Part of the International Conference of Agricultural Economists*, Milan, Italy, August 9–14.

Chibwana, V., Fisher, M., & Shively, G. (2011). Land allocation effects of agricultural input subsidies in Malawi. *World Development, 40*(1), 124–133.

Coady, D. P. (1995). An empirical analysis of fertilizer use in Pakistan. *Econometrica, 62*(246), 213–224.

Croppenstedt, A., Demeke, M., & Meschi, M. M. (2003). Technology adoption in the presence of constraints: The case of fertilizer demand in Ethiopia. *Review of Development Economics, 7,* 58–70.

Dorward, A., Kydd, J., Morrison, J., & Urey, I. (2004). A policy agenda for pro-poor agricultural growth. *World Development, 32*(1), 73–89.

Dorward, A., Chirwa, E., Kelly, V., Jayne, T. S., Slater, R., & Boughton, D. (2008). *Evaluation of the 2006/07 agricultural input subsidy programme, Malawi* (Final Report). Lilongwe, Malawi.

Duflo, E., Kremer, M., & Robinson, J. (2008). How high are rates of return to fertilizer? Evidence from field experiments in Kenya. *American Economic Review, 98*(2), 482–488.

Duflo, E., Kremer, M., & Robinson, J. (2011). Nudging farmers to use fertilizer: Theory and experimental evidence from Kenya. *American Economic Review, 101*(6), 2350–2390.

Ellis, F., & Maliro, D. (2013). Fertiliser subsidies and social cash transfers as complementary or competing instruments for reducing vulnerability to hunger: The case of Malawi. *Development Policy Review, 31*(5), 575–596.

FAO (Food and Agriculture Organization). (2000). *Fertilizers and their use: A pocket guide for extension officers* (4th ed.). Rome and Paris: Food and Agriculture Organization of the United Nations and International Fertilizer Association.

Fisher, M., & Kandiwa, V. (2014). Can agricultural input subsidies reduce the gender gap in modern maize adoption? Evidence from Malawi. *Food Policy, 45,* 101–111.

Gregory, D. I., & Bumb, B. L. (2006). *Factors affecting supply of fertilizer in Sub-Saharan Africa* (Agriculture and Rural Development Discussion Paper 24). Washington, DC: World Bank.

Holden, S., & Lunduka, R. (2010). *Too poor to be efficient? Impacts of the targeted fertilizer subsidy program in Malawi on farm plot level input use, crop choice and land productivity.* Unpublished manuscript. Ås, Norway: Department of Economics and Resource Management, Norwegian University of Life Sciences.

Holden, S., & Lunduka, R. (2012). Do fertilizer subsidies crowd out organic manures? The case of Malawi. *Agricultural Economics, 43*(3), 303–314.

Jack, B. K. (2013). Private information and the allocation of land use subsidies in Malawi. *American Economic Journal: Applied Economics, 5,* 113–135.

Jayne, T. S., & Rashid, S. (2013). Input subsidy programs in Sub-Saharan Africa: A synthesis of recent evidence. *Agricultural Economics, 44,* 547–562.

Jones, C., & Jacobsen, J. (2003). Fertilizer placement and timing. MSU extension service. In *Nutrient management module 11*. East Lansing, MI: Michigan State University.

Kilic, T., Whitney, E., & Winters, P. (2015). Decentralized beneficiary targeting in large-scale development programs: Insights from the Malawi farm input subsidy program. *Journal of African Economies, 24*(1), 26–56.

Lunduka, R., Ricker-Gilbert, J., & Fisher, M. (2013). The consistent and contrasting results of research examining the economic impacts of Malawi's fertilizer subsidy program. *Agricultural Economics, 44*(6), 563–579.

Marenya, P., & Barrett, C. B. (2009). State-conditional fertilizer yield response on western Kenyan farms. *American Journal of Agricultural Economics, 91*(4), 991–1006.

Matsumoto, T., & Yamano, T. (2011). Optimal fertilizer use on maize production in East Africa. In T. Yamano, K. Otsuka, & F. Place (Eds.), *Emerging development of agriculture in East Africa: Markets, soil, and innovations*. The Netherlands: Springer.

Mazvimavi, K., Murendo, C., Minde, I., & Kunzekweguta, M. (2013). *Assessing the impacts of Zimbabwe's agricultural vouchers input program*. Paper presented at the African Association of Agricultural Economists Meeting in Hammemet, Tunisia, September 22–25.

Minot, N., & Benson, T. (2009). *Fertilizer subsidies in Africa: Are vouchers the answer?* (Issue Brief 60). Washington, DC: International Food Policy Research Institute.

Minten, B., Koru, B., & Stifel, D. (2013). The last mile(s) in modern input distribution: Pricing, profitability, and adoption. *Agricultural Economics, 44*(6), 629–646.

Niang, I., Ruppel, O. C., Abdrabo, M. A., Essel, A., Lennard, C., Padgham, J., & Urquhart, P. (2014). Africa. In V. R. Barros, C. B. Field, D. J. Dokken, M. D. Mastrandrea, K. J. Mach, T. E. Bilir, et al. (Eds.), *Climate change 2014: Impacts, adaptation, and vulnerability. Part B: Regional aspects. Contribution of working group II to the fifth assessment report of the intergovernmental panel on climate change* (pp. 1199–1265). Cambridge, United Kingdom and New York, NY: Cambridge University Press.

Nsengiyumva, F., Peralta-Alva, A., Maehle, N., Ruiz, S. L., Wu, F., Banda, S., et al. (2015). *Malawi selected issues* (International Monetary Fund Country Report No. 15/346). Washington, DC.

Pan, L., & Christiaensen, L. (2012). Who is vouching for the input voucher? Decentralized targeting and elite capture in Tanzania. *World Development, 40*(8), 1619–1633.

Ricker-Gilbert, J., & Jayne, T. S. (2012). *Do fertilizer subsidies boost staple crop production and reduce poverty across the distribution of smallholders in Africa? Quantile regression results from Malawi*. Selected paper for the Triennial Meeting of the International Association of Agricultural Economists, Foz Do Iguacu, Brazil, August 18–24, 2012.

Ricker-Gilbert, J., & Jayne, T. S. (2017). Estimating the enduring effects of fertilizer subsidy programs on recipient farm households. Evidence from Malawi. *Journal of Agricultural Economics* (Forthcoming).

Ricker-Gilbert, J., Jayne, T. S., & Chirwa, E. (2011). Subsidies and crowding out: A double-hurdle model of fertilizer demand in Malawi. *American Journal of Agricultural Economics, 93*(1), 26–42.

Sheahan, M., & Barrett, C. B. (2014). *Understanding the agricultural input landscape in Sub-Saharan Africa. Recent plot, household, and community-level evidence* (World Bank Policy Research Working Paper 7014). Washington, DC.

Sheahan, M., Black, R., & Jayne, T. (2013). Are Kenyan farmers under-utilizing fertilizer? Implications for input intensification strategies and research. *Food Policy, 41*, 39–52.

Snapp, S. S. (1998). Soil nutrient status of smallholder farms in Malawi. *Communications in Soil Science and Plant Analysis, 29*(17 and 18), 2571–2588.

Snapp, S., Jayne, T. S., Mhango, W., Benson, T., & Ricker-Gilbert, J. (2014). *Maize yield response to nitrogen in Malawi's smallholder production system* (Malawi Strategy Support Program. Working Paper 9). Washington, DC: International Food Policy Research Institute.

Tsoka, M. G., Mezuwa, H., Mbendera, M., & Nyirongo, I. (2015). *Evaluation of FISP E-voucher pilot programme 2013/14*. Report prepared by the Centre for Social Research University of Malawi, Zomba.

Vanlauwe, B., Kihara, J., Chivenge, P., Pypers, P., Coe, R., & Six, J. (2011). Agronomic use efficiency of N fertilizer in maize-based systems in Sub-Saharan Africa within the context of integrated soil fertility management. *Plant and Soil, 339*, 35–50.

Xu, Z., Guan, Z., Jayne, T. S., & Black, R. (2009). Factors influencing the profitability of fertilizer use on maize in Zambia. *Agricultural Economics, 40*(4), 437–446.

2

Smallholder Farmers' Access to Inputs and Finance in Africa

Augustine Langyintuo

Abstract Agriculture remains the mainstay of most economies in Africa, accounting for 37% of gross domestic product (GDP), nearly 60% of export earnings, and over 76% of employment. Over the decades, agricultural value added has averaged 17%, growing at a rate of 1.4%. This has been driven primarily by low use of modern technologies such as improved crop varieties, fertilisers and other complementary inputs, which in turn is an artefact of limited access to agricultural finance. The direct consequence of low agricultural productivity is the high rates of poverty and malnutrition on the continent. African governments are renewing efforts to promote agriculture, recognising that GDP growth originating in agriculture is about four times more effective in reducing poverty than GDP growth in other sectors, although how best to do this within very complex political economies remains a challenge. As an important first step towards reducing poverty and increasing wealth among smallholder farmers in Africa, this chapter explores the major challenges in farmers' access to productive farm inputs and finances and reviews alternative approaches that could be used to improve access by farmers to these resources.

1 Introduction

Agriculture remains the mainstay of the economies of many African countries. Nearly 60% of export earnings are from agriculture, and over 76% of the 987 million Africans living in rural areas are employed in agriculture (FAO 2010). Although agriculture's average contribution to GDP on the continent has been declining over time, it still remains high, averaging 37% with a range of 3–67%. Whereas agricultural value added averages 17%, value added per agricultural worker in 2003 (in 2000 USD) averaged USD 327, growing at a rate of 1.4%, compared to USD 23081 at a rate of 4.4% in Organisation for Economic Cooperation and Development (OECD) countries during the same period (IAASTD 2009).

A. Langyintuo (✉)
International Finance Corporation, World Bank Group, Delta Center, Menengai Road, PO Box 30577, Upper Hill, Nairobi, Kenya
e-mail: alangyintuo@ifc.org

In Africa, agricultural productivity growth over the decades has been disappointingly low. Observed increases in cereal production have been due, primarily, to increase in cultivated area (Fig. 1). By contrast, in Asia in the 1960s, production increases resulted from the rapid uptake of high-yielding wheat and rice varieties and the use of fertilisers and irrigation combined with subsidies, which drove down the cost of production and raised land and labour productivities (Hazell and Ramasamy 1991). The observed increases in production still fall short of population growth, compelling African governments to import cereals in the order of 50 million megatonnes, at an estimated cost of USD 30–50 billion a year. It is estimated that if continental food supplies do not increase, Africa will spend about USD 150 billion on food imports by 2030 (IFPRI 2012).

Numerous factors account for the low productivity of agriculture in Africa, not least the limited use of improved agricultural technologies, especially improved seeds, fertilisers and mechanisation services, which in turn is an artefact of the lack of access to agricultural finance. Additionally, the apparent lack of land tenure security is thought to be a hindrance to increased investment in land improvement technologies and to encourage the adoption of unsustainable agricultural practices, leading to poor family incomes and nutrition (UNECA 2005). The consequence of the low productivity in the predominantly agro-based economies on the continent is pervasive poverty. More than half of the extreme poor live in SSA, with around 413 million people living on less than USD 1.90 per day (Fig. 2) (World Bank 2013, 2018); a staggering 249 million are undernourished (FAO 2019).

Several experts have observed that the expansion of smallholder farming can lead to a faster rate of poverty alleviation, by raising the incomes of rural cultivators and reducing food expenditure, thus reducing income inequality (Magingxa and Kamara 2003; Diao and Hazell 2004; Resnick 2004; Barham and Chitemi 2008; World Bank

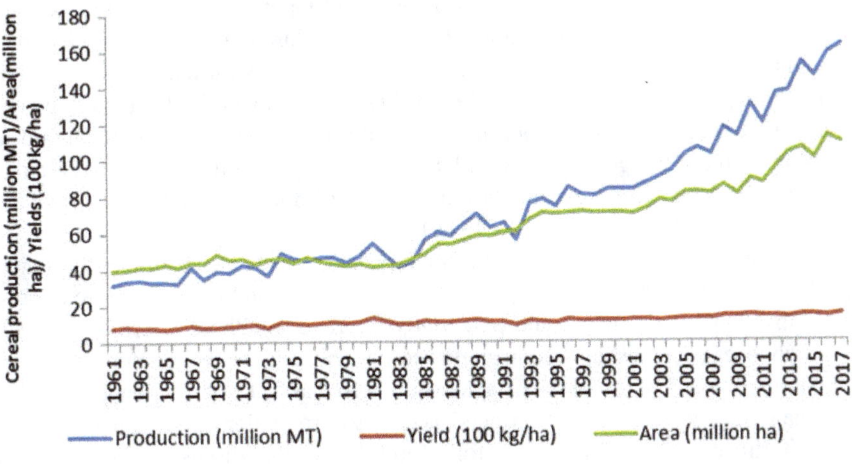

Fig. 1 Trends in cereal production, area planted and productivity in sub-Saharan Africa (SSA). *Source* FAO (2018)

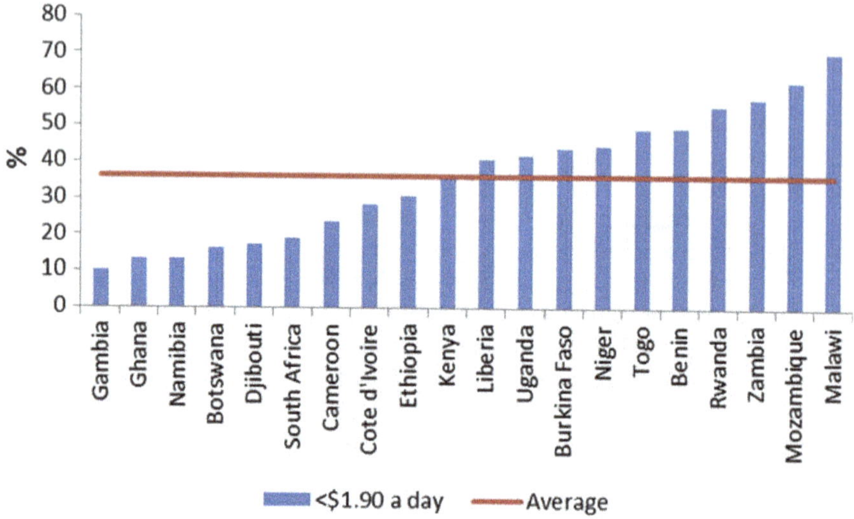

Fig. 2 Population living below the poverty line of USD 1.90/day (%). *Source* World Bank (2018)

2008a). This is consistent with the 2008 World Development Report, which pointed out that GDP growth originating in agriculture is about four times more effective in reducing poverty than GDP growth in other sectors (World Bank 2008a). A 10% increase in productivity can reduce poverty by 4% in the short run and 19% in the long run (FAO 2010).

In the light of this, most African governments embrace reduction of poverty and increase of wealth among smallholder farmers as important policy challenges. This is exemplified by the African Heads of State 2003 Maputo Comprehensive African Agricultural Development Programme (CAADP) declaration, which commits governments to spend at least 10% of their national budgets on agriculture to raise agricultural productivity to at least 6% (NEPAD 2004); and the 2006 Abuja Declaration that 'given the strategic importance of fertiliser in achieving the African Green Revolution to end hunger, the African Union Member States resolve to increase the level of use of fertiliser from the current average of 8 kilograms per hectare to an average of at least 50 kg per hectare by 2015'.

Drawing on existing literature and secondary data, this chapter examines the major challenges to farmers' access to productive farm inputs and finances and explores alternative approaches that could potentially improve smallholder farmers' access to productive resources, as a contribution to government efforts to improve lots of rural households. The rest of the Chapter is organised as follows: Section 2 examines the challenges to smallholder farmers' access to production inputs, mainly land, seeds, fertiliser and finance. This is followed by a detailed discussion on strategies to improve smallholder farmers' access to inputs and finance in Sect. 3. Section 4 presents the concluding remarks on the chapter.

2 Challenges to Smallholder Farmers' Access to Production Inputs

Farmers typically need various resources to organise agricultural production, but for the sake of this chapter we focus on purchased inputs (seeds and fertiliser) and access to finance. We acknowledge upfront that access to secure land is equally important, deserving a separate chapter, and it is therefore only briefly discussed here to contextualise the discussion.

2.1 Limitations in Access to Secure Agricultural Land

Africa accounts for over 60% of the available arable land on Earth. Nevertheless, smallholder farmers in Africa are unable to secure sufficient and suitable land to grow their crops and keep livestock. As shown in Fig. 3, land pressure is severe in Rwanda, Malawi, Kenya and Uganda, with an average holding of barely one hectare, compared to Burkina Faso, Mali, Niger and Ghana, where the average holding is more than three hectares. Not only is access a problem, the security of access is an even bigger problem facing farmers. This is partly because of the predominantly customary land tenure system observed in many countries including Mali, Zambia, Malawi, Ghana, Burkina Faso and Niger and in large parts of Sierra Leone, Liberia, Nigeria, Tanzania and Mozambique. Under such tenure arrangements, land tends to be held collectively by lineages or families without providing any form of security to users, especially women and young people (Namubiru-Mwaura et al. 2012). In most parts of Africa,

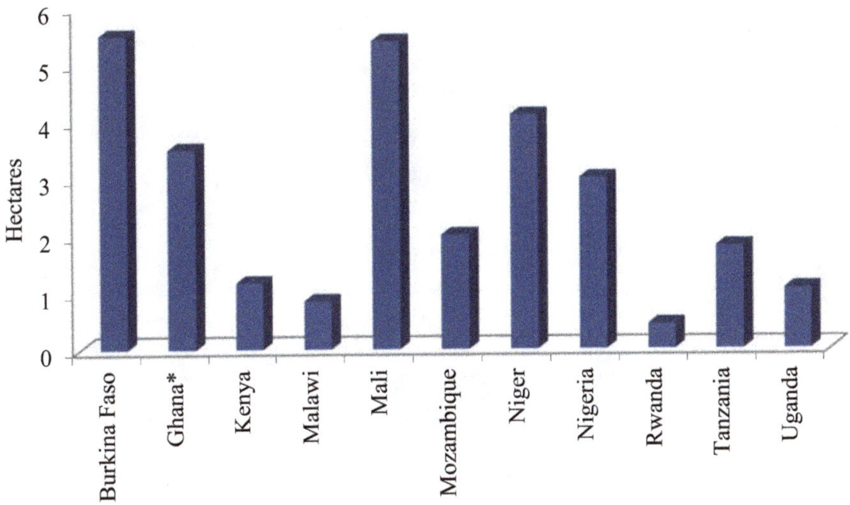

Fig. 3 Average area of agricultural land per household (2009–2010). *Source* AGRA (2014)

women's rights to land are limited to 1–2% of land and dependent on their marital statuses, although evidence suggests they contribute more than 70% of agricultural labour (Bennett 2010). The problem of land tenure insecurity is exacerbated by state interference, through acquisitions and forceful seizure of farmlands in the name of investment.

Tenure security affects agricultural productivity through the choice of crop to grow, limited investment in land and adoption of unsustainable agricultural practices (UNECA 2005). Although privatisation of land would seem to be effective in reducing insecurity, evidence seems to suggest that although short-term land rentals improve land productivity (Kebede 2002; Holden et al. 2008), they provide no incentives for either the landlord or the tenant to make long-term improvements (Place 2009), thereby compromising on sustainable production.

Land productivity is largely influenced by access to reliable water sources, especially under predominantly rainfed conditions often characterised by significant climatic variability. About 60% of SSA is exposed to drought, and 30% extremely (Hodson et al. 2009), yet irrigation facilities are limited (Fig. 4). Most of the existing irrigation facilities are ineffectively and inefficiently utilised. This is because constructions are often fraught with problems, such as generally insufficient farmer involvement in design; development often far removed from existing farming systems; inadequate land tenure system development for irrigation; capital-intensive investment requiring high input levels; and chronic institutional weaknesses.

The Intergovernmental Panel on Climate Change 2007 Report predicts that up to 250 million people in Africa will experience problems in accessing sufficient water by 2020 because of climate change, potentially leading to halving of agricultural

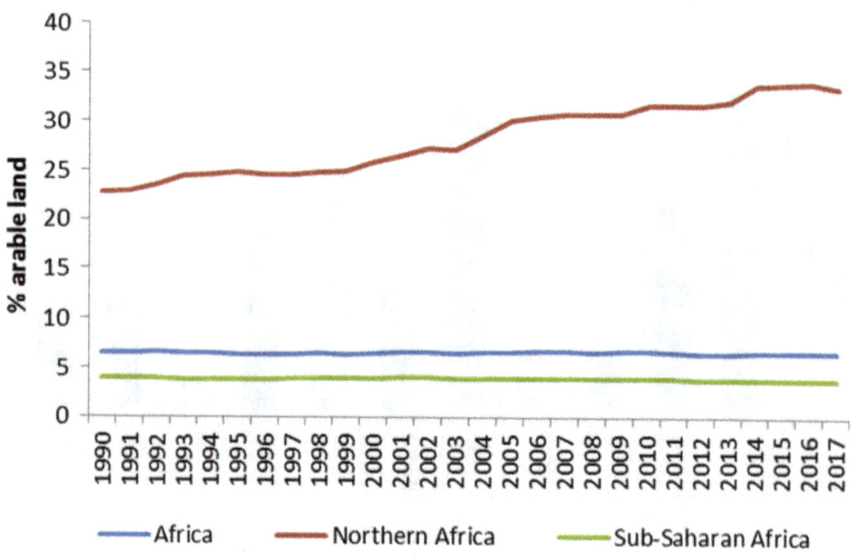

Fig. 4 Percentage of arable land equipped with irrigation. *Source* FAO (2019)

production (IFPRI 2012). The report also predicts that, without adaptation, the impact of climate change on agriculture and food security will be high, with the number of malnourished children possibly increasing by an extra 10 million to a total of 52 million by 2050.

2.2 *Policies and Institutional Factors Hindering Farmers' Access to Improved Seed*

Seed is an essential, strategic and relatively inexpensive input to agriculture, with a high rate of return on investment that often sets the upper limit for crop production. Improved seeds typically yield 4–6 mt/ha, compared to traditional unimproved ones that yield less than 1 mt/ha. Yet the uptake of the former is limited in Africa. Langyintuo et al. (2010) suggest that the adoption rates for improved maize[1] seed average 28% of the cultivated area of approximately 17 million ha (Table 1). This low adoption rate is blamed on a combination of policy and technical problems that hinder the supply of and demand for improved seeds. On the demand side, risk aversion among farmers, lack of knowledge of the availability of ecologically adapted varieties, relatively high seed price and lack of cash resources are the main determinants of adoption. The high-risk aversion observed among smallholder farmers is due

Table 1 Adoption rate of improved maize varieties in selected countries in Africa

Country	Area (million ha)	Seed demand (1000 t)	Adoption rate (% of area)
Ethiopia	1.7	42	19
Kenya	1.6	39	72
Tanzania	2.6	64	18
Uganda	0.7	17	35
Angola	0.8	19	5
Malawi	1.4	35	22
Mozambique	1.2	30	11
Zambia	0.6	14	73
Zimbabwe	1.4	34	80
Benin	0.7	16	Na
Ghana	0.7	19	1
Mali	0.3	8	0.3
Nigeria	3.6	89	5
Total/average	17.3	427	28

Source Langyintuo et al. (2010)

[1]Maize is one of the most important food crops grown in Africa.

to the absence of physical assets, which diminishes their risk-bearing abilities, and hence, their reluctance to invest in untried technologies, including improved seed.

The lack of knowledge of adaptable varieties is primarily due to weak extension service delivery, relative to the numerous unfamiliar varieties released onto the market without adequate farmer education on the types and economic benefits of improved varieties, to improve their adoption decisions. Unfortunately, extension coverage is weak and sometimes skewed towards the relatively richer farmers (Langyintuo and Setimela 2007). Farmer confidence in the improved seed is sometimes further eroded by the proliferation of fake seeds on the market.

Some unscrupulous traders engage in unethical advertising practices, or simply painting grains in colours similar to known and trusted genuine varieties, to undercut prices. This not only cheats farmers out of their meagre cash resources, but permanently damages the loyalty built over time.

The relatively high seed prices are the combined effects of market policy failures and supply-side imperfections (discussed below). Whereas market imperfections sometimes cause misalignment of seed and grain prices, policy failures often lead to high production and marketing costs, poor seed quality assurance and uncompetitive seed markets leading to inferior pricing mechanisms ultimately affecting farmers negatively. For example, policymakers often attempt to improve consumer welfare by imposing price ceilings on outputs, as part of their market reforms strategy, without any attempt to make similar adjustments to seed prices. The end result is that farmers, who are less organised, are forced to buy seed at relatively high prices, thereby subsidising urban consumers to prevent urban unrest at the expense of their own welfare. Although free seed handouts by governments and non-governmental organisations are designed to address the liquidity constraints of farmers, they are known to have negative impacts on rural seed market development, as beneficiaries tend to be unwilling participants in the commercial seed market.

Although many countries have made significant progress in liberalising and restructuring their seed sectors in the past two decades (Hassan et al. 2001), some still operate dated seed policies or none at all, partly contributing to the incidence of fake seeds (Langyintuo 2004). Where policies exist, they almost exclusively concentrate on the formal seed sector and fail to support the diversity of initiatives that farmers employ for their seed security (Louwaars and Engels 2008). In most cases, the emphasis is always on hybrids to the neglect of open-pollinated varieties, as observed in India by Spielman et al. (2009).

Even where there are updated policies, their implementation may sometimes pose a significant challenge to seed sector development. For instance, most national governments in Africa insist on the registration of all newly developed varieties, to ensure the genetic identity of the variety and discourage the release of germplasm that is inappropriate, unproductive or unsafe. However, the registration processes have been observed in many countries to be very lengthy (up to three or more years) and expensive. Depending on the country, a breeder may pay between USD 1000 and USD 2500 per entry per year for both national performance trials (NPTs) and the distinctness, uniformity and stability (DUS) test, which are necessary components of the registration process (Langyintuo et al. 2010; Mwala and Gisselquist 2012).

Not only are these costs ultimately passed on to farmers, but the process lengthens the time it takes farmers to access newly developed varieties.

Regional spillover of genetic improvement, through harmonisation of regional seed laws, can significantly reduce the costs of seed development and shorten the time it takes for farmers to benefit from improved genetics. Unfortunately, this has become problematic because the legislative frameworks of countries within regional economic communities vary widely in facilitating harmonisation. For example, plant variety protection is not enforced in countries such as Angola, Malawi, Uganda and all West African countries excluding Ghana. Ethiopia and Uganda are yet to update their Seeds Acts, while International Seed Testing Association (ISTA[2]) and OECD accreditation required for official seed shipment across borders are available in only Ghana, Kenya, Malawi, South Africa, Zambia and Zimbabwe. Differences in certification systems, standards and procedures have led to diminished trust among seed certification authorities in the different countries.

It is important to comment on the use of biotechnology in crop genetic improvement. For many years, biotechnology has been providing value-added foods and medicines for mankind. Recent advances in genomics, including the ability to insert genes across species, have distinguished 'modern biotechnology' from traditional methods. Resulting transgenic or genetically modified (GM) crops, forestry products, livestock and fish have potentially favourable qualities such as pest and disease resistance, however, with possible risks to biodiversity and human health (Paarlberg 2014). With the exception of four African countries (Table 2), the use of GM varieties remains controversial, largely driven by negative perceptions originating from Western consumers and exported to Africa (De Groote et al. 2014; Clive 2012; Paarlberg 2000, 2002, 2008). It is important to point out that GM crops have been subject to more testing worldwide than any other new crops and have been declared as safe as conventionally bred crops by scientific and food safety authorities worldwide (Paarlberg 2014). As noted by Paarlberg (2014), a recent EU report concludes that more

Table 2 Area planted to GM crops in Africa in 2012 (ha)

Country	Crop planted			Total
	Cotton	Soybean	Maize	
Burkina Faso	300,000	0	0	300,000
Egypt	0	0	1000	1000
South Africa	15,000	362,000	1873,000	2300,000
Sudan	200,000	0	0	200,000
Total	515,000	382,000	1874,000	2801,000

Source Compiled from Clive (2012)

[2]A country without an ISTA-accredited laboratory, such as Angola, Ethiopia, Mozambique, Tanzania or Uganda, cannot sell seed across borders, since any cross-border traded seed lots must bear ISTA Seed Lot and Sample Certificates, to certify that the seed has met the requirements of the ISTA rules.

than 130 EU research projects, covering a period of more than 25 years of research and involving more than 500 independent research groups, concur that consuming foods containing ingredients derived from GM crops is no riskier than consuming the same foods containing ingredients from conventional crops. Such well-known organisations as the World Health Organization, the US National Academy of Sciences and the European Food Safety Authority (EFSA) have come to the same conclusion (Paarlberg 2014).

2.3 *Constraints to Smallholder Farmers' Access to Fertiliser in Africa*

Organic and inorganic (or mineral) fertilisers are strategic inputs to crop production, especially where the existing soils are exhausted from continuous cropping without adequate soil amelioration. Evidence shows that about 25% of crop production is lost each year without application of nitrogen fertiliser; by the 10th year, 60% is lost (Donovan and Casey 1998). Nonetheless, the average consumption of inorganic fertilisers is very low, at around 16 kg/ha of nutrients—ranging from less than 1 kg/ha in Niger and Gambia to about 89 kg/ha in South Africa (Fig. 5). This is compared with 331 kg/ha in East Asia and Pacific and 160 kg/ha in South Asia and over 180 kg/ha in the upper middle income world (World Bank 2019). Within SSA, Zambia, South Africa and Côte d'Ivoire have achieved the target in the Abuja Declaration of 50 kg/ha. Whereas high levels of fertiliser use create environmental problems in developed countries and in a few countries in Africa such as Egypt and

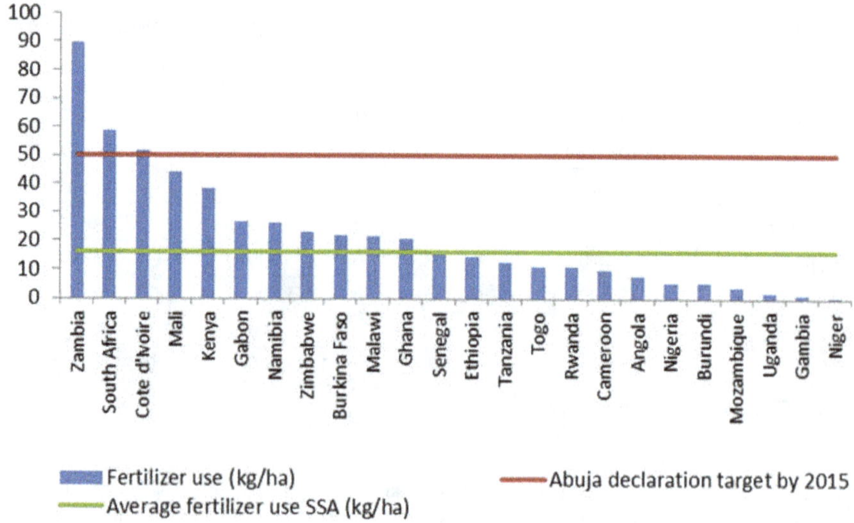

Fig. 5 Intensity of fertiliser use in selected countries in Africa. *Source* World Bank (2019)

Morocco, in most parts of Africa, the limited use of the input creates environmental degradation leading to an estimated loss of 4–12% of GDP, through soil mining and clearing of forest land to expand farms in an attempt to increase production (Olson and Berry 2003).

It is believed that demand and supply-side policy failures are to blame for the limited use of fertilisers in Africa. On the demand side, the risk of fertiliser use and the poor nitrogen to maize price ratio that has been trending downward by 0.9% are a disincentive to fertiliser use (Gregory and Bumb 2006; Heisey and Norton 2007; Morris et al. 2011). Because most of the crops grown by farmers are staples and non-tradable while fertilisers are imported, currency devaluation often increases the price of fertiliser several times above output prices.

In addition, fertiliser prices are uncompetitive because of the slow emergence of the private sector and consequent lack of a vibrant market, which in turn is an artefact of unfavourable private-sector policies: poorly defined rules of the game, weak regulatory enforcement, proliferation of taxes and fees, cumbersome bureaucratic importation procedures, general lack of security and widespread incidence of corruption (World Bank 2006). Prices are further increased by poor road infrastructure and the cost of finance. Added to the many official and unofficial tolls and taxes, security check points along the roads slow the delivery of services and impose transaction costs.

Unfortunately, the high cost of importation and distribution of fertiliser is likely to remain for a long time to come. At the present level of African fertiliser markets development it is cost-effective to import until markets expand to support large-scale local production (World Bank 2006). Presently, over 90% of the fertiliser used in Africa is imported at very high sourcing costs, which ultimately reduce the profitability of distributing fertiliser and discourage increased supply. The scope for negotiating bulk purchases and arranging bulk shipments in order to save on freight charges is limited by the lack of port facilities capable of handling large volumes.[3]

2.4 Access to Agricultural Finance for Smallholder Farmers

Agriculture is the predominant activity in African economies, yet less than 4% of total commercial bank lending goes into the agricultural sector (Fig. 6). Financial institutions often cite lack of usable collateral, high transaction costs due to remoteness of clients, dispersed demand for financial services, the lag between investment needs and expected revenues, lack of irrigation, pests and diseases, small size of farms and of individual transactions, underdeveloped communication and transportation infrastructure and high covariate risks due to variable rainfall and price risks (Adesina

[3]Most of the fertiliser imported into Africa is shipped via 10,000 tonne vessels because of limited capacities at the ports, especially those outside of South Africa. This limits the size of bulk orders and entails a shipping cost premium of 10–15% over medium-sized vessels (Morris et al. 2011). All these factors negatively affect farm gate prices, thereby constraining the use of fertiliser by smallholder farmers.

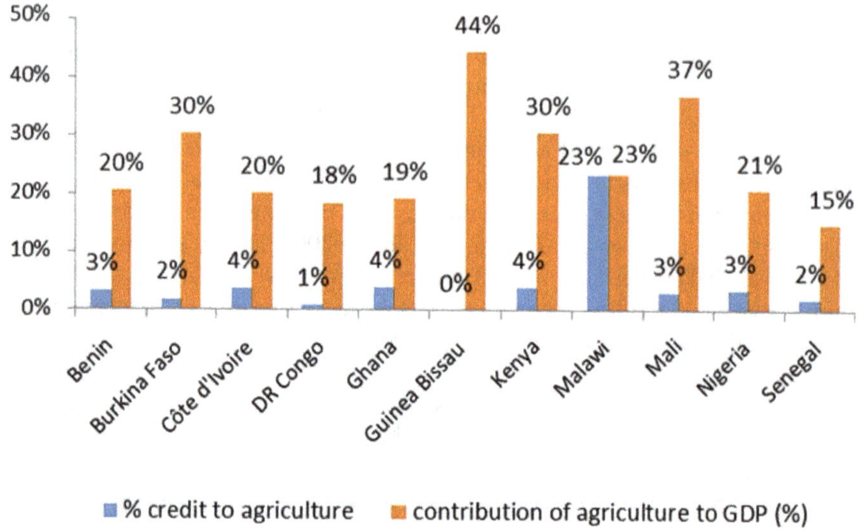

Fig. 6 Agricultural lending as a share of agricultural GDP in selected African countries. *Source* FAO (2018)

et al. 2012) as reasons why they do not lend to smallholder farmers. Other challenges include poorly developed agri-food value chains, which significantly increase risks and exposure for the bank, and general lack of understanding among financial institutions of the agricultural sector and the opportunities.

In principle, the unsatisfied demand by smallholder farmers and SMEs for financial services can be met by microfinance institutions (MFIs). These institutions have emerged to provide credit facilities and deposits but have not succeeded in expanding financing for agriculture, due to a number of reasons including limited capital bases, high interest rate, small size of disbursement insufficient for investment and being located in urban centres when the bulk of farmers are in rural areas. Moreover, the repayment schedules for microfinance loans often do not synchronise with the seasonality of agriculture and the timing of farmers' cash flows.

Furthermore, Poulton et al. (2006) noted that some of the challenges faced by MFIs included very small outreach compared to demand, inadequate capacity to properly conduct credit analysis and loan appraisals, inadequate risk management and control systems, small shares of total deposits and loans compared to commercial banks, non-performing collateral laws limiting the effectiveness of the MFIs and inadequate capitalisation limiting levels when assessing the potential of default by a prospective borrower.

3 Strategies to Improve Smallholder Farmers' Access to Inputs and Finance

3.1 *Improving Security of Land Access*

In recent times, there has been some progress in the development of land policy frameworks in Africa, but a recent report (FAO 2010) showed that many of them are weak in addressing ethnic and gender issues, land information systems and monitoring mechanisms. This is possibly because, under customary systems, the land is usually accessed through complex social relations governed by local institutions, and hence, national land policies and laws often have little relevance. Therefore, any policy reform must be tailored to the physical, social and economic contexts, as well as taking into consideration economic factors, equity issues and less tangible concerns such as the social or religious beliefs that people attach to land. The framework should also consider the capacity of the country to implement such policies.

3.2 *Facilitating Access to Improved Seeds*

It is apparent from the analysis that adoption rates are low and farmers' propensity to buy seed can be enhanced in a number of ways. Firstly, farmers need to know about the existence, characteristics and economic value of a given variety, through the dissemination of timely extension messages. This seems to be lacking due to weak extension systems. To address this problem, some seed companies invest in extension message delivery, for instance by simply printing symbols on the seed packs that depict the maturity group of the variety. For example, SeedCo Limited uses the image of an elephant for a long maturing variety and a zebra for an early maturing variety.

Secondly, farmers need credit themselves to be able to purchase the seed. Partly due to the non-competitive nature of the seed industry, seeds are generally priced above the means of farmers and their participation in the seed market would require support through targeted subsidies or subsidised farm credit.

Thirdly, farmers deserve better returns on their investment in seeds to encourage them to continue to invest. Efforts to make output value chains profitable are critical in enhancing seed demand.

Finally, governments need to pursue some key reforms to improve the supply side. These include updating seed legislature to be consistent with the development of the sector and to be private sector-friendly to encourage private investment. The public sector should demonstrate a willingness to domesticate harmonised regional seed laws, regulations and standards to promote regional seed trade. Given the widespread faking of seeds in the region, legislation should provide for stiffer punishment for those convicted of the offence, as a deterrent to others.

In terms of biosafety, countries need a functional legislative framework encompassing the relevant policies, laws, regulations and requisite administrative structures and processes to safely harness GM technology. It is acknowledged that most African countries are at various stages of creating enabling environments for GM crop commercialisation. Five countries (Cameroon, Kenya, Malawi, Nigeria and Uganda) are currently conducting field trials of biotech crops, the final step before full approval for commercialisation.

3.3 *Promoting Fertiliser Access for Farmers*

To encourage farmers to use fertilisers, the practice of using blanket fertiliser recommendations, which are sometimes sub-optimal in specific situations, should be discouraged in favour of ecology/crop-specific rates. There should be an emphasis on the use of micronutrients and on demonstrating the profitability of fertiliser use. Strategies that enhance fertiliser use, such as microdosing and organic/inorganic fertiliser combinations, should also be considered. While appreciating the economic burden on governments and development partners, it is important to provide smart subsidies to farmers to promote fertiliser demand (Jayne and Jones 1997; Kelly et al. 2003).

On the supply side, government regulations are required to ensure competitive supply chains, which at present are generally weak in many countries. Efforts should be made to prevent importers and wholesalers from collusive practices, including price fixing and market segmentation.

Public policies should address fertiliser sourcing[4] costs and distribution costs, the availability and cost of business finance and risk management instruments. Adequacy of supply chain coordination mechanisms could improve the fertiliser value chain. Other areas of policy intervention include access to foreign exchange and credit, and strengthening port infrastructure. Expanding capacity in the main ports of entry to allow larger vessels to discharge can help reduce the landed cost of fertiliser.

3.4 *Facilitating Agricultural Finance for Smallholder Farmers*

The financing gap in agriculture created by the commercial banks may be closed by exploring various options, including credit guarantees, interlocked markets for finance and warehouse receipts systems.

---

[4] Sourcing costs could be reduced through pooling import orders.

3.4.1 Credit Guarantee Schemes

To address the perception of high risk and lack of collateral limiting commercial banks' lending to farmers, various development practitioners have rolled out innovative financing approaches, including credit guarantee schemes over the past few years. Credit guarantees have been used to cover part of the default risk, ensuring secure repayment of all or part of the loan in case of default (Levitsky 1997). Besides covering the default risk, credit guarantees are useful in addressing the issue of lack of collateral and poor credit history faced by farmers, and hence improve loan terms. Additionally, allowing loans to be made to borrowers who would otherwise have been excluded from the lending market enables farmers and SMEs to establish a repayment reputation in future (De Gobbi 2002) and benefit from lower transaction costs and helps to raise productivity (Ruiz Navajas 2001; Green 2003).

The use of credit guarantee schemes must be guided by best practice (World Bank 2008), as discussed here. Firstly, whether the scheme should focus on individual or portfolio loans is important. In an individual loan arrangement, the application is approved by the guarantor and the application is assessed on a case by case basis, thereby establishing a direct link between the borrower and the lender. An estimated 72% of credit guarantee schemes use this selective or individual loan approach (World Bank 2008b). While allowing for more careful risk management and likely reducing the probability of moral hazard, this approach introduces a high cost of loan management.

Secondly, the fees charged for the use of credit guarantees have a direct impact on the incentives for lenders and borrowers in participating in the scheme, as well as on the financial sustainability of the fund. Although it is not realistic to expect credit guarantees to cover full costs through fees, the fees must be high enough to cover administrative costs, but low enough to ensure adequate lender and borrower participation.

Thirdly, the default rate is an important indication of the sustainability of a guarantee scheme. A sustainable scheme should aim to have a default rate of 2–3%, thanks to a critical assessment of the application and effective monitoring. Newly established schemes in developing countries might consider a higher default rate (i.e. over 5%) in their early years of operation, but should aim at lower rates in the shortest possible time. In general, guarantee payouts should only be used if all efforts by the guarantors to reschedule payments have failed.

Fourthly, the risk-sharing arrangement between guarantor, lender and borrower defines the efficiency and effectiveness of the guarantee scheme. An improperly designed guarantee scheme can increase moral hazard among borrowers by reducing the default risk they would otherwise face, while a properly designed guarantee scheme can limit moral hazard. The guarantor should accept enough risk to be able to persuade banks to participate in the scheme, while reducing the scope for moral hazard or adverse selection. The level of risk sharing depends on which part of the agricultural value chain the scheme intends to focus on. For lending towards the upper part of the agricultural value chain—agro-processors, agro-dealers, fertiliser and seed companies, etc.—direct risk sharing at 50:50 would be sufficient, as the risk of lending is lower. For lending to the lower part of the agricultural value chain, especially to poor smallholder farmers, higher levels of risk-sharing arrangements such as first loss arrangements will be required. In general, a risk-sharing rate below 50% reduces the potential for moral hazard but tends to reduce the incentives for banks to participate in the guarantee scheme, because of high loan administration costs (World Bank 2008b).

Lastly, guarantee schemes should consider using risk management mechanisms such as reinsurance, loan sales or portfolio securitisation, in order to reduce the exposure to default and diversify risk. Globally, about 76% of credit guarantee schemes use risk management tools, 20% loan insurance, 10% securitised loans portfolio and 5% risk management strategies (World Bank 2008b). It should, however, be noted that these mechanisms require relatively well-developed local capital and financial markets.

3.4.2 Interlocked Markets for Credit and Value Chain Financing

One way of overcoming missing markets in the supply of credit to farmers for purchasing improved seeds and fertilisers is to use value chain or interlocked markets for inputs, outputs and credit (Poulton et al. 1998). Traditionally done through government-controlled parastatal agencies (Poulton et al. 1998), this type of value chain financing—including contract farming, vertically integrated operations or outgrower schemes—has often been viewed as the major source of credit for farmers (IFAD 2003). This has often been led by the private sector and dominated by agro-processors, agribusinesses and traders, and food-processing companies—either local or international—operating under international markets have emerged (Swinnen and Maertens 2010). These value chain financing arrangements are becoming increasingly important for farmers, especially in export-oriented value chains such as horticulture, as a source of finance for inputs and markets for the outputs. For example, farmers growing cash crops such as cotton sometimes receive fertilisers for these crops but also use the inputs on their food crops, inevitably increasing and sustaining cash and food crop production (Dione 1991).

There are many advantages to interlocked credit market arrangements, not least the low risk of default because the cost of inputs is deducted before the farmers receive payments for the produce delivered. It also offers assured markets, guaranteed prices, reduced marketing risks and sharing of lending risks with positive spillover effects on other crops (Swinnen and Maertens 2010). Monitoring and supervision costs are also reduced since the input loan is delivered in kind to farmers to be applied on their crops, except when farmers engage in side selling of the produce. This practice is particularly difficult to curb where there are no appropriate legal frameworks to enforce contracts.

3.4.3 Warehouse Receipts System (WRS)

WRS is becoming an important instrument for accessing financing from commercial banks. As noted earlier, one of the reasons why banks are reluctant to lend to agriculture is the lack of collateral for loans. The absence of land titles and other encumbered fixed assets that could be used as collateral, as well as the lack of indemnity for product quality deterioration and storage losses, diminishes banks' willingness to lend to the agricultural sector. A WRS can potentially be used to unlock the collateral value of inventories that farmers, traders and processors manage through warehouse receipt financing, helping to relieve some of the existing constraints to accessing credit.

A producer, farmer group, trader, exporter, processor or indeed any individual or body corporate may deposit grains or other commodities into a registered warehouse. The warehouse operator issues the depositor with a warehouse receipt (WR),[5] which stipulates the quality, quantity and type of commodity, the date deposited and the date up to which storage costs have been paid (Swinnen and Maertens 2010). The warehouse operator holds the stored commodity by way of safe custody, implying he is legally liable to make good at any value lost through theft or damage by fire and other catastrophes, but has no legal or beneficial interest in it.

There are many benefits to the WRS, including easing rural finance, reducing storage losses, improving grades and standards, facilitating trade, enhancing market efficiency, mitigating price risks and enabling cost-effective management of public food reserves (Coulter and Onumah 2002). Focusing on the easing of rural finance, the depositor can take the warehouse receipt to a bank or other financial institution as collateral for a short-term loan, usually up to 60–70% of the value of the crop. Availability risk associated with movable collateral is reduced by the warehouse operator's guarantee of delivery from a stated location, and foreclosure can be simple and low cost without any resort to the courts, depending on the legal regime (Coulter and Onumah 2002). The risk of loss of value of the collateral can be minimised by monitoring movements in its market value and using margining and price risk management instruments. Lenders no longer need to monitor a large number of small borrowers, just a few warehouse operators, to assure loan performance. This ultimately reduces monitoring costs and encourages commercial lending to the rural sector, helping to capitalise the rural trade.

4 Concluding Remarks

Agriculture is the mainstay of the economies of African countries but productivity growth has been disappointingly low, perpetuating hunger and poverty. The main

[5]The receipts may be transferable, allowing transfer to a new holder—a lender (where the stored commodity is pledged as security for a loan) or a trade counterparty—which entitles the holder to take delivery of the commodity upon presentation of the WR at the warehouse.

reasons for the poor performance of the agricultural sector have been a combination of low use of improved agricultural technologies (mainly seeds and fertiliser) and dysfunctional production and marketing policies. To reverse the trend, priority support to the sector should focus on the implementation and enforcement of predictable private-sector-friendly laws and regulations. Opening up market opportunities for inputs and outputs, through the enforcement of regional harmonisation, would increase the spillover impacts of technologies, thereby lowering the cost of inputs. Additionally, priority areas to promote agricultural transformation in Africa should include the development of financial policies that are coherent and private-sector-friendly and implementing risk-sharing instruments to leverage commercial bank credit into agriculture to capitalise the sector.

References

Adesina, A. A., Langyintuo, A., Bugo, N., Makinde, K., Bigirwa, G., & Wakiumu, J. (2012). Improving farmers' access to agricultural inputs and finance: Approaches and lessons from sub-Saharan Africa. In Hazell et al. (Ed.), *New directions for smallholder agriculture*. Oxford: Oxford University Press.

AGRA (Alliance for a Green Revolution in Africa). (2014). African agriculture status report. AGRA, Nairobi, Kenya.

Barham, J., & Chitemi, C. (2008). *Collective action initiatives to improve marketing performance lessons for farms groups in Tanzania: Collective action for property rights (CAPRI)*. Working paper No. 74.

Bennett, P. (2010). Women's rights and access to land in Africa. New Agriculturist.

Clive, J. (2012). *Global status of commercialized biotech/GM Crops: 2012* (ISAAA Brief No. 44). ISAAA: Ithaca, NY.

Coulter, J., & Onumah, G. (2002). The role of warehouse receipt systems in enhanced commodity marketing and rural livelihoods in Africa. *Food Policy, 27*, 319–337.

De Gobbi, M. (2002). *Making social capital work: Mutual guarantee associations for artisans, social finance programme, employment sector*. International Labour Organisation, September.

De Groote, H., Kimenju, S. C., Keter, F., Ngigi, O., & Gitonga, Z. (2014). *But what do rural consumers in Africa think about GM food?* Paper Prepared for presentation at the 18th ICABR Conference, Nairobi, June 2014.

Diao, X., & Hazell, P. (2004). Exploring market opportunities for African smallholders. In *2020 Conference Brief No. 6*. Washington D.C., USA: International Food and Policy Institute.

Dione, J. (1991). Food security and policy reform in Mali and the Sahel. In P. Dasgupta (Ed.) *Issues in Contemporary Economics: Proceedings of the Ninth World Congress of the International Economic Association* (Vol. 3). London: Macmillan).

Donovan, G., & Casey, F. (1998). *Soil fertility management in sub-Saharan Africa* (60 p) (World Bank technical paper No. 408). Washington DC, USA: World Bank.

FAO (Food and Agriculture Organization of the United Nations). (2010). *Africa's changing landscape: Securing land access for the rural poor*. Rome, Italy. Available at http://www.fao.org/docrep/012/al209e/al209e00.pdf.

FAO (Food and Agriculture Organization of the United Nations). (2018). *Credit to agriculture data*. Available at: http://www.fao.org/economic/ess/investment/credit/en/.

FAO. (2019). *The state of food security and nutrition in the world. Safeguarding against economic slowdowns and downturns*. Rome, Italy: Rome. Food and Agriculture Organization of the United Nations. Available at http://www.fao.org/state-of-food-security-nutrition/en/.

Green, A. (2003). *Credit guarantee schemes for small enterprises: An effective instrument to promote private sector-led growth?* The United Nations Industrial Development Organization (UNIDO). Working Paper No. 10, August.

Gregory, D. I., & Bumb, B. L. (2006). *Factors affecting supply of sub-Saharan fertilizer in Africa* (Agriculture and rural development discussion paper 24). Washington, DC: World Bank.

Hassan, R. M., Mekuria, M., & Mwangi, W. M. (2001). *Maize breeding research in eastern and southern Africa: Current status and impacts of past investments by the public and private sectors, 1966–1997*. Mexico, D. F. (Mexico): CIMMYT.

Hazell, P. B. R., & Ramasamy, C. (1991). *The green revolution reconsidered: The impact of high yield rice varieties in South India* (Food Policy Statement Number 14. International Food Policy Research Institute) Washington, DC. USA.

Heisey, P. W., & Norton, G. (2007). Fertilizer and other farm chemicals. In R. E. Evenson, P. Pingali, & T. P. Schultz (Eds.), *Handbook of agricultural economics. Agricultural development: Farmers, farm production, and farm markets* (Vol. 1, 3A). Amsterdam, The Netherlands: Elsevier.

Hodson, D. P., Cressman, K., Nazansi, K., Park, R. F., & Yahahoui, A. (2009). The global cereal monitoring system. In *Abstract for the Borlaug global rust initiative technical workshop*, 17–20 March, Cindad, Obregon, Mexico. P21. Available at: www.globalrust.org. Last Accessed on April 24, 2009.

Holden, S., Otsuka, K., & Place, F. (Eds.). (2008). *The emergence of land markets in Africa: Impacts on poverty, equity, and efficiency*. London: Routledge.

IAASTD (International Assessment of Agricultural Knowledge, Science and Technology for Development). (2009). *Agriculture at a crossroads* (Sub-Saharan Africa (SSA) Report, Vol. V). International Assessment of Agricultural Knowledge, Science and Technology for Development.

IFAD (International Fund for Agricultural Development). (2003). *Agricultural marketing companies as sources of smallholder credit in east and southern Africa: Experiences, insights and potential donor roles*. Rome, Italy: IFAD East and Southern Africa Division, IFAD.

IFPRI (International Food Policy Research Institute). (2012). Increasing Agricultural productivity & enhancing food security in Africa. In *Synopsis of an International Conference New Challenges Opportunities*. Washington, DC. USA: International Food Policy Research Institute.

Jayne, T. S., & Jones, S. (1997). Food marketing and pricing policy in Eastern and Southern Africa: A survey. *World Development, 25*(9), 1505–1527.

Kebede, B. (2002). Land tenure system and common pool resources in rural Ethiopia: A study based o fifteen states. *Africa Development Review, 14*, 113–150.

Kelly, V., Adesina, A., & Gordon, A. (2003). Expanding access to agricultural inputs in Africa: A review of recent market development experience. *Food Policy 28*, 379404.

Langyintuo, A. S. (2004). *Challenges of the seed sector of southern Africa* (p 39) (International Maize and Wheat Improvement Centre (CIMMYT Report). Harare, Zimbabwe. At www.cimmyt.org/gis/rfseedsafrica/.

Langyintuo, A. S., & Setimela, P. (2007). *Assessment of the effectiveness of maize seed assistance to vulnerable farm households in Zimbabwe*. Mexico, D.F.: CIMMYT.

Langyintuo, A. S., Mwangi, W., Diallo, A. O., MacRobert, J., Dixon, J., & Bänziger, M. (2010). Challenges of the maize seed industry in eastern and southern Africa: A compelling case for private–public intervention to promote growth. *Food Policy, 35*, 323–331.

Levitsky, J. (1997, February, May). Best practice in credit guarantee schemes. *The Financier*, Vol. 4, No. 1 and 2.

Louwaars, N. P., & Engels, J. M. M. (2008). *Seed policies: Enabling support to informal seed systems*. Research Gate. Available on http://www.researchgate.net/publications/37789145.

Magingxa, L., & Kamara, A. (2003). *Institutional perspectives of enhancing smallholder market access in South Africa*. Paper presented at the 41st Annual Conference of the Agricultural Economic Association of South Africa held in Pretoria. Diao and Hazell 2004.

Morris, M., Ronchi, R., & Rohrbach, D. (2011). Building sustainable fertilizer markets in Africa. In *Proceedings of International Conference on Priority Actions Toward Market Development for African Farmers*. Nairobi, Kenya: International Livestock Research Institute.

Mwala, M., & Gisselquist, D. (2012). *Private-sector agricultural research and innovation in Zambia: Overview, impact, and policy options*. Lusaka, Zambia: University of Zambia. Technology Foundation (AATF), Nairobi, Kenya.

Namubiru-Mwaura, E. L., Knox, A., & Hughes, A. (2012). *Customary land tenure in Liberia: Findings and implications drawn from 11 case studies* (Report prepared for the Liberia Land Policy & Institutional Support (LPIS) Project). Washington, DC: United States Agency for International Development.

NEPAD. (2004). *Implementing the comprehensive Africa agriculture development programme and restoring food security in Africa*. Available at www.nepad.org/2005/files/documents/172.pdf. NEPAD, South Africa.

Olson, J., & Berry, L. (2003). *Land degradation in Uganda: Its extent and impact*. Mimeo.

Paarlberg, R. L. (2000). *Governing the GM crop revolution policy choices for developing countries* (p. 44) (Food, Agriculture, and the Environment Discussion Paper 33). Washington, D. C: International Food Policy Research Institute.

Paarlberg, R. L. (2002). The real threat to GM crops in poor countries: Consumer and policy resistance to GM foods in rich countries. *Food Policy, 27,* 247–250.

Paarlberg, R. (2008). *Starved for science: How biotechnology is being kept out of Africa*. Cambridge, MA: Harvard University Press.

Paarlberg, R. L. (2014). *Why green revolution food crops went forward, while GMO food crops did not.* http://www.fao.org/fileadmin/templates/cfs_high_level_forum/documents/REPORT_HLEF.pdf 2014.

Place, F. (2009). Land tenure and agricultural productivity in Africa: A comparative analysis of the economics literature and recent policy strategies and reforms. *World Development, 37,* 1326–1336.

Poulton, C., Kydd, J., & Dorward, A. (2006). Overcoming market constraints on pro-poor agricultural growth in sub-Saharan Africa. *Development Policy Review, 24*(3), 243–277.

Poulton, C., Dorward, A., & Kydd, J. (1998). The rival of smallholder cash crops in Africa: Public and private roles in the provision of finance. *Journal of International Development, 10*(1), 85–103.

Resnick, D. (2004). *Smallholder African agriculture: Progress and problems in confronting hunger and poverty* (DSGD Discussion Paper No. 9). Washington, D.C: International Food Policy Research Institute.

Ruiz Navajas, A. (2001). Credit guarantee schemes: Conceptual frame. Financial system development project, GTZ/FONDESIF, November.

Spielman, D., Ekboir, J., & Davis, K. (2009). The art and science of innovations systems inquiry: Applications to sub-Saharan African Agriculture. *Technology in Society, 31,* 399–405.

Swinnen, J. F. M., & Maertens, M. (2010). *Finance through food and commodity value chains in globalized.* Paper presented at the KFW Financial Sector Development Symposium Finance for Food: Towards new agriculture and rural finance.

UNECA (Unite Nations Economic Commission for Africa). (2005). *Structural transformation to breakaway from poverty. Chap. 4 in Economic report on West Africa.* Available at http://www.UNECA.org/eva2005/chap4.pdf. UNECA, Addis Ababa, Ethiopia.

World Bank. (2006). *Promoting increased fertilizer use in Africa: Lessons learned and good practice guidelines* (Africa Fertilizer Strategy Assessment ESW Technical Report Discussion Draft). Washington, DC. USA.

World Bank. (2008a). *The growth report: Strategies for sustained growth and inclusive development.* Washington D.C.: Commission on Growth and Development, World Bank.

World Bank. (2008b). *The typology of partial credit guarantee funds around the world* (Policy Research Working Paper 4771. Development Research Group). Washington DC: The World Bank.

World Bank. (2013). *Poverty headcount ratio at national poverty lines (% of population).* Available at: http://data.worldbank.org/indicator/SI.POV.NAHC. Last Accessed on March 2013.

World Bank. (2018). *Poverty and shared prosperity 2018: Piecing together the poverty puzzle.* Washington, D.C.: The World Bank.

World Bank. (2019). *The World Bank data.* Available at: https://data.worldbank.org/indicator/AG.CON.FERT.ZS?view=chart.

3

Impact of Casual and Permanent Off-Farm Activities on Food Security: The Case of India

Alwin Dsouza, Ashok K. Mishra and Tadashi Sonoda

Abstract India is the largest producer of food grains, dairy commodities and horticultural crops and largest exporter of rice, beef and cotton. But the country, at the micro-level, still struggles with extensive and deep-rooted problems with food security. The Planning Commission notes that 22% of the 1.2 billion Indians are still living in poverty. This chapter assesses the impact of off-farm income and labour allocation (both casual and permanent off-farm work), and decisions on food security of smallholder households in India. Promoting food distribution schemes and female education would bring about further reduction of food insecurity among rural households. We found that off-farm activities and off-farm business income reduce food insecurity of rural Indian households. Spouse's casual off-farm work status had a negative impact on household's food insecurity—increased food security; operator's casual off-farm work status had a positive impact on household's food insecurity—increased food insecurity. However, food insecurity of households increased if both operator and spouse worked casually off the farm.

1 Introduction

Traditionally, farmers in developing countries were involved in farming full-time and rarely opted for non-farm work. However, with significant growth in the rural economy in many developing countries, the rural non-farm sector has experienced a substantial increase in the share of rural employment. Additionally, variability in climate, riskiness in farm income, credit market constraints and shortage of hired labour have also led to income diversification strategies (farm and off-farm) by farmers, both in the developed and developing countries (Ruben and Berg 2001; Ellis 2000; Mishra and Goodwin 1997). As a result, smallholder farmers have changed their labour allocations, both on and off the farm. Additionally, it can be argued that off-farm work

A. Dsouza · A. K. Mishra (✉)
Arizona State University, Mesa, AZ, USA
e-mail: Ashok.K.Mishra@asu.edu

T. Sonoda
Nagoya University, Nagoya, Japan

could increase the educational outcomes and productivity of other family members[1] (Yang 1997). Finally, the literature reveals the significance of off-farm income in increasing food security for rural households. For example, Owusu et al. (2011) concluded that non-farm work positively influenced the food security status of farm households in the northern region of Ghana. In another study, Barrett et al. (2001) conclude that off-farm work could be a possible pathway out of the vicious circle of poverty for rural households.

However, the literature falls short of differentiating the type of off-farm job (casual/part-time or permanent/full-time[2]). Specifically, the literature fails to discuss the impacts of types of off-farm income and labour allocation decisions, by the operator and spouse, on food security in smallholder households in developing countries. Therefore, the objective of this chapter is to assess the impact of off-farm income and labour allocation, both casual and permanent, on the food security of smallholder households in India. Additionally, we attempt to understand the dynamics of the above impacts when either the operator or the spouse—or both—is involved in off-farm activities. Our study further contributes to the literature by using panel data which have an advantage over cross-sectional data. We use wider, nationally representative household-level data than previously reported. Finally, we compare the mean impact from the panel data with heterogeneous impact from quantile panel regressions.

The chapter is organised as follows. Section 2 discusses off-farm work and food security in India, with a special focus on the status of off-farm work—permanent and/or casual. This section also reviews the literature on the existing role of operator–spouse relationship on food security. Section 3 presents the data used in the empirical analysis. Section 4 outlines the conceptual framework and empirical procedure. Section 5 discusses empirical results, followed by concluding remarks.

2 Food Security and Off-Farm Work in India

India, located in South Asia, is a fast-growing country. Its population is growing by 2% a year (current population of 1.24 billion). However, 69% of Indians live in rural areas and 58% of these rural households are agricultural households—directly dependent on agriculture for their employment and livelihood (NSSO 2104). The agricultural sector's contribution to gross domestic product (GDP) fell from 43 to 14% over the period 1970–2011. Rapid growth in non-agricultural sectors (service and manufacturing sectors) has led to the migration of labour out of agriculture. Nonetheless, agriculture is an important sector of the economy, accounting for around

[1]Recall that off-farm work has been carried out by an educated member of the smallholder household and in turn, other members of the family either start going to school or receive better information and farming experience.

[2]In the case of Israel, Kimhi (1994) found that government incentives rewarded farmers to engage in full-time farm work. Additionally, in the case of the US, Goodwin and Mishra (2004) found that off-farm work decreased farming efficiency.

14% of GDP and 11% of the country's exports. About 56% of the land mass is agricultural land, and only 43% is net cultivated area; only about 45% of cropped area is reported to be irrigated. The net cultivated area increased significantly, by about 18%, from 119 million hectares in 1950–51 to about 140 million hectares in 1970–71. Since then, net cultivated area has remained stable at around 140 million hectares; only 3.5% of the area is under permanent crops. However, the average holding size is about 1.3 ha, and about 85% of farms fall into this category (known as smallholders).

Although India is the largest producer of food grains, dairy commodities and horticultural crops, and the largest exporter of rice, beef and cotton, the country, at the micro-level, still struggles with extensive and deep-rooted problems with food security. The Planning Commission notes that 22% of the 1.2 billion Indians are still living in poverty (Planning Commission 2014). Furthermore, the per capita income of Indians is about 15% that of the world overall, and one in three children is malnourished. In response to the 2007–2008 global food crisis, the Indian government enacted the National Food Security Mission (NFSM) which resulted in a significant increase in food grain production. All indicators show that the NFSM has helped to attain India's food security at the national level. However, at the micro-level, the situation is characterised by extensive malnutrition and stunting among the population. As a result, the government enacted the National Food Security Act (NFSA) in 2013. The NFSA 2013 is seen as a vital step in alleviating the issue of widespread poverty and malnutrition. But the majority of poor households either reside in rural areas or are dependent on agriculture. Therefore, any policy measures designed to lift the poor out of poverty may include advancements in the agricultural sector.

The Indian rural labour market is mainly characterised by two types of work status: regular (or permanent) and casual work, defined on the basis of payment method. For example, permanent work is where wages are paid at monthly or regular intervals, sometimes accompanied by payments in advance or bonuses/gifts during the harvest season. On the other hand, casual work is where wage payments are made either daily or on a piecemeal basis; casual labourers could be hired in groups (Pal 1996). Finally, women in rural areas are generally excluded from regular (or permanent) jobs and tend to hold more casual jobs. The choice of work status depends on wage and another non-wage criteria, including credit constraints and home production time, especially for women. Given poor working conditions and long working hours on the farm, casual jobs are preferred to farm work. Casual work is very common in the farming sector. However, the opposite is true in the case of the rural non-farm sector[3]. In the rural non-farm sector, both casual and permanent jobs coexist, but permanent jobs are relatively scarce. Permanent jobs are generally held by the rich and educated class (upper caste), while poor and uneducated workers are engaged in casual non-farm jobs (Lanjouw and Shariff 2004). On the other hand, readily available, casual non-farm jobs can also increase agricultural wages, leading to an increase in the incomes of rural poor who are dependent on farming activities. For example, the

[3]The rural non-farm sector is defined as economic activities in rural areas other than agriculture, livestock, fishing and hunting (Lanjouw and Lanjouw 2001).

successful implementation of public works programmes such as the Employment Guarantee Scheme and *Jawahar Rozgar Yojana* (employment generation scheme) has led to an increase in demand for casual labour in the non-farm sector, leaving a shortage of hired labour in the agricultural sector (Berg et al. 2012; Kareemulla et al. 2010; Bhargava 2014). The rural non-farm sector has become a dominant source of employment and income for smallholder households in India, consequently affecting the food security of smallholder households.

It should be recalled that women in rural areas tend to work as casual labourers. The majority of women in rural areas work as a cultivator or a casual labourer on other land (Lanjouw and Murgai 2009). If that is the case, then income-generating farm and off-farm activities may have implications for the food security of smallholder households. It is widely known that, compared to men, women spend a higher proportion of their income on food, child health, education[4] and other consumption items (Quisumbing et al. 1995; Thomas 1991). Quisumbing et al. (1995) further state that, relative to men, women's incomes are comparatively less in amount and more in frequency. Women spend the majority of their income on daily consumption, while men spend more on durable or expensive goods. In another study, Thomas (1991) found that, compared to paternal income, maternal income had 4–8 times greater impact on family health. This also holds true for child survival probabilities, which are 20 times higher. In this study, we argue that a wage shock (casual or permanent wages) would affect the food security of smallholder households. For example, a shock in the casual wage of the casual worker, in the rural non-farm sector, is likely to have a significant impact on the food security, compared to a shock in the regular wage of the permanent worker. Similarly, the impact of a drop in casual and permanent wages differs depending on the gender of the head of the rural smallholder household. Finally, the working status (casual or permanent off-farm work) of women/spouses could have a substantial impact on the food security of smallholder households. Our study covers the impact of the work status of either operator or spouse, or both, on the food security status of the household.

3 Data

Data for this study are drawn from two rounds of Indian Human Development Survey (IHDS) held in 2004/05 and 2011/12, respectively. IHDS is a nationally representative survey of about 41,554 households and 215,754 individuals, in 1503 villages and 971 urban areas of India. In the second round, around 83% of the household were re-interviewed; around 2134 new households were also interviewed in the second round. Information about split households was also considered. The survey included information regarding household income, consumption and standard of living; for individuals, it included information on employment, morbidity and education.

[4] In a recent study Afridi et al. (2015) concluded that the mother's participation in the labour force would improve the educational outcomes of the child.

In this study, we considered a panel of 17,142 rural households that were interviewed in both rounds. Only rural households were chosen for this study, because we are interested in analysing the impact of work status (casual and permanent) on the food security of smallholder households. In this study, the share of food expenditure is the sum of food expenditure over total expenditure. Food expenditure is the sum of expenditures on various food items, including food at restaurants, consumed in the last 30 days, and total expenditure is the total monthly household expenditure[5] which includes health, rent, communication, transport, education, clothing and footwear. The share of food expenditure over total household expenditure in our study is considered as an indicator of food insecurity status (Smith and Subadoro 2007). Returning to our measure of food insecurity, an increase in the share of food expenditure over total expenditure would make the household more food insecure. This is because, if the household experiences a negative income shock, then it would be accompanied by a reduction in food consumption, making the household more food insecure (Smith and Subadoro 2007). The opposite is also true.

The socio-demographic explanatory variables include age of household head and spouse, years of education of both household head and spouse, size of household, number of older persons (≥64 years) living in the household, number of children (≤14 years) living in the family, whether the household receives subsidised food (below poverty line—BPL cardholder), and wealth index[6]. Along with these, shares of farm income, off-farm income, and off-farm business income[7] are included in the model. To further analyse the dynamics within the household, a casual and permanent form of off-farm work is also considered. Casual work is defined as work that is undertaken daily, casual piecemeal work, or any contract lasting less than a year while permanent work is defined as work under a regular term or longer contract. Finally, smallholder households where the operator, spouse or both are involved in casual and permanent off-farm work are considered in this study.

4 Descriptive Statistics

Table 1 shows a considerable change in the scenario for Indian rural households in recent years. The share of food expenditures fell considerably, from 57% in 2004/05 to 49% in 2011/12. Average food expenditures representing a lower proportion of total expenditures may indicate that significant populations in rural areas are becoming less food insecure. The average age of operator and spouse was around 45 and

[5]Only few expenses were reported for a reference period of 365 days; these were then converted to monthly expenses by dividing by 12.

[6]Wealth index derived by using principal component analysis of various common assets owned by the households. This included ownership of house, bicycle, scooter/motorcycle, television, cooler, electric fan, telephone, mobile phones, refrigerator, car, air conditioner, washing machine, computer, laptop, credit card and microwave oven.

[7]Farmers may have off-farm businesses (such as a seed, tractor or implement company, tea stall or restaurant) that provide income to the household.

Table 1 Variable definition and descriptive statistics, rural households, India, 2004/05 and 2011/12

Variables	2004/05	2011/12	Average
Share of food expenditure (%)	57	49***	53.02
Age of head of household (years)	45.23	51.41***	48.32
Age of spouse (years)	39.74	46.81***	43.24
Educational level of head of household (years)	1.51	1.58**	1.55
Educational level of spouse (years)	2.35	2.42*	2.39
Household size (number)	5.59	5.24***	5.42
Number of family members ≥64	0.18	0.39***	0.29
Number of family members ≤14	1.57	0.92***	1.25
Share of farm income[a] (%)	18.05	16.14***	17.10
Share of off-farm income[b] (%)	49.72	50.22	49.97
Share of business income[c] (%)	9.37	8.15***	8.76
Wealth index[d]	−0.0001	−0.0002	0.0001
Below poverty line[e] (BPL) cardholders (%)	40.11	48.21***	44.16
Only operator casual[f] off-farm worker (%)	28.43	22.65***	25.54
Only operator permanent[g] off-farm worker (%)	6.4	5.54***	0.8
Only spouse casual off-farm worker (%)	2.9	4.73***	3.8
Only spouse permanent off-farm worker (%)	0.3	0.6***	0.5
Both casual off-farm workers (%)	23.47	25.16***	24.32
Both permanent off-farm workers (%)	0.45	0.64**	0.5
Spouse permanent; operator casual off-farm worker (%)	0.30	0.70***	0.51
Operator permanent; spouse casual off-farm worker (%)	0.80	0.90	0.87
No of observations	17,251	17,251	17,251

***$p < 0.01$, **$p < 0.05$, *$p < 0.10$

Source Indian Human Development Surveys (IHDS), 2004/05 and 2011/12

[a]Includes incomes from agricultural property, crop residue, crop and livestock receipts

[b]Includes incomes from wages and salaries from off-farm agricultural work and off-farm non-agricultural (non-farm) work. These can also be classified into casual and permanent work

[c]Includes incomes from off-farm businesses that the household owns. These businesses include big business, selling cloth, food such as pickles, selling in the market, skilled services such as a doctor, or unskilled service such as a barber

[d]Includes ownership of house, bicycle, sewing machine, generator set, mixer/grinder, scooter/motorcycle, black/white television, colour television, air cooler, clock/watch, electric fan, telephone, chair/table, cot, mobile phone, telephone, refrigerator, pressure cooker, car, air conditioner, washing machine, computer, laptop, credit card and microwave oven

[e]Defined as rural households living below poverty line (INR 447 and INR 816 per capita per month, for 2004/05 and 2011/12, respectively.)

[f]Defined as when operator/spouse engaged in casual daily work, casual piecemeal work or contractual work for less than one year

[g]Defined as when operator/spouse engaged in longer contracts of more than one year

40 years, respectively, in 2005. The average level of education (in years) is higher for the operator. The size of household averaged around five members in both years. The number of older members (≥64 years) living in the household, on average, was significantly lower in both years. On average, every rural household had at least one child aged under 14. This highlights the fact that rural India is significantly young. Farm income contributed, on average, around 16–18% in both years, while non-farm income, was on average a major contributor to household income.

It should be recalled that off-farm income, a key variable in our study, is defined as income from non-farm activities other than farming income (own farm). Off-farm income may also include income from work performed on other farms and/or work performed in other rural non-farm sectors, such as construction, manufacturing or the services sector. Table 1 reveals that off-farm income, as a share of total smallholder household income, has been steady—around 50% in both survey years (2004/05 and 2011/12). Smallholders in India could also derive income by engaging in non-farm business income. Off-farm businesses include a seed company, a farm implement dealership, storage facilities and produce stands. Table 1 shows that business income was around 8–9% of total household income in the survey years. Interestingly, Table 1 reveals that the share of people living below the poverty line (BPL cardholders)[8] has increased from 40% in 2004/05 to 48% in 2011/12. This increase in the share of BPL cardholders may be attributed to increased awareness by rural households of social security programmes such as the public distribution system (PDS) for basic food items.

Another set of variables germane to our study is the work status of the operator and spouse. For example, Table 1 shows that the share of the workforce (operators or spouses) employed permanently in off-farm jobs is significantly low. In all, 6.0–6.5% of rural households have reported either the operator or spouse working off the farm permanently. Additionally, data from 2004/05 and 2011/12 in Table 1 show a declining trend in operators' engagement in permanent jobs in the off-farm labour market. However, Table 1 shows that the share of spouses working permanently in the non-farm sector has increased moderately, from 0.3% in 2004/05 to 0.6% in 2011/12. On the other hand, the share of spouses involved in casual off-farm jobs has increased significantly, from 2.9% in 2004/05 to 4.8% in 2011/12.

Our study also includes variables where both operator and spouse were engaged in permanent off-farm jobs and/or where both were engaged in casual off-farm jobs. Such time allocation decisions may be important because differing work status (casual or permanent) may have different implications on the income, capacity to withstand production/consumption shocks and food security status of smallholder households. Table 1 reveals that around 24–25% of smallholder households have both operator and spouse engaged in casual off-farm jobs; about 0.50–0.60% of smallholder households report both working in permanent off-farm jobs. Finally, less than 1% of the smallholder households report either operator with permanent off-farm job and spouse with casual off-farm job, or vice versa.

---

[8] Defined as rural households living below poverty line (INR 447 and INR 816 per capita per month in 2004/05 and 2011/12 respectively).

Let us now compare the sources of income for smallholder households, by the households' food insecurity (FIS) class[9] (see Smith and Subandaro 2007), for 2004/05 and 2011/12 (Table 2). In 2005, households in the least FIS class derived about 16% of their total income from farming, while the income from farming for households in the highest FIS class was around 20% (column 5, Table 2). However, by 2011/12 the share of farm income in total income for these classes decreased to 14% and 16%, respectively. Findings here highlight the riskiness in farm income associated with variability in output and/or prices in the farming sector. Interestingly, the share of off-farm income in total household income is consistent across all FIS classes. For example, regardless of FIS class, off-farm income contributed about 48–51% towards total household income in 2004/05 and 2011/12. This is consistent with the fact that off-farm income is stable and has little or no variability. Table 2 reveals a considerable difference in the share of business income to total income. For instance, the least FIS class received about 13% of their total income from off-farm business ownership. However, for the highest FIS class, off-farm business income was about 6% of total income. A similar trend is also observed in 2011/12 (Table 2, lower panel).

A consistent trend across all FIS classes is sources of income diversification (Table 2). We observe a higher reliance on off-farm income and income from off-farm businesses. Along with this, we also observe a significant share of household income from other sources, including remittance, property income and other public benefits (Table 2). Table 2, lower panel, shows that in 2011/12 the *least FIS class* derived almost 2 5% of their total income from other sources. Additionally, regardless of FIS class, the share of other income in total income has increased over time (2004/05–2011/12). This finding underscores the importance of other sources of income in overcoming food insecurity in rural India. Table 2 also shows that over time, the proportion of rural households in the high and highest FIS classes has fallen considerably—30 to 25% in the high FIS class and 35–20% in the highest FIS class. This is a positive trend, which also corroborates with the poverty level in rural India, as estimated by the Planning Commission of India, for the respective years.[10] In the next section, we present the methodology that estimates the effect of off-farm income and labour allocation decisions by operators and spouses on food insecurity, using both panel analysis and quantile regression with fixed effects.

[9]Based on Smith and Subandaro (2007), we identify four FIS classes. The class is based on the share of food expenditure as a proportion of total household expenditure. Least food insecure (FIS) class: <=50%; medium FIS class: >=50%– <=65%; high FIS class: >=65%–<75%; highest FIS class: >=75%.

[10]The poverty ratio for rural India was 41.8% in 2004/05, falling to 25.7% in 2011/12 (Planning Commission of India 2014).

Table 2 Sources of income, by share of food expenditures classification, 2004/05 and 2011/12

Year	Source	Share of food expenditure classification (%)				
		<=50% least food insecure class	gt;=50%–65% medium food insecure class	>=65%–75% high food insecure class	>=75% highest food insecure class	Average
2004/5	Farm income[a] share (%)	16	17	18	20	18
	Off-farm income[b] share (%)	48	49	51	49	50
	Business income[c] share (%)	13	11	9	6	9
	Other income[d] share (%)	23	23	22	25	23
	Households (%)	15	20	30	35	
2011/12	Farm income[a] share (%)	14	16	16	18	16
	Off-farm income[b] share (%)	50	51	51	49	50
	Business income[c] share (%)	11	9	7	5	8
	Other income[d] share (%)	25	24	26	28	26
	Households (%)	20	35	25	20	

Source Calculated from Indian Human Development Surveys (IHDS), 2004/05 and 2011/12

[a]Includes income from agricultural property, crop residue, crop and livestock receipts

[b]Includes income from wages and salaries from off-farm agricultural work and off-farm non-agricultural (non-farm) work. These can also be classified into casual and permanent work

[c]Includes income from off-farm businesses that the household owns. These businesses include big business, selling cloth, food such as pickles, selling in the market, skilled services such as a doctor, or unskilled service such as a barber

[d]Includes income from remittances, property and other income such as rent earned, interest and dividends, government/private pensions, other income from public benefits including scholarships, gifts, national old age pension scheme, widows' pension scheme, national maternity scheme, national disability pension, Annapurna, other government transfers, NGOs or other assistance

5 Methodology

In order to estimate the effect of off-farm income and labour allocation for both operator and spouse on food insecurity, we use mean effects and heterogeneous effects modelling.

5.1 *Estimation of Mean Impact*

The mean impact of off-farm income and labour allocation decisions on food insecurity is estimated using the following model:

$$Y_{ijt} = \beta E_{ijt} + \delta X_{ijt} + \alpha_{ij} + \vartheta_{jt} + \theta_t + \varepsilon_{it}$$

where β and δ are the coefficients to be estimated, α_{ij} is a set of time-constant unobservable household variables, ϑ_{jt} is a set of time-variant unobservable village variables, θ_t is a year dummy (=0 for 2004/05 and =1 for 2011/12) and ε_{it} is a set of time-variant unobservable household variables. The dependent variable (Y_{ijt}) is the share of food expenditure in total consumption expenditures or FIS. E_{ijt} is a vector of main variables of interest and X_{ijt} is other explanatory variables. We control for age and education of both operator and spouse, wealth index, BPL cardholders, household size, number of older members ($\geq$64 years) in the household and children ($\leq$14 years) in the household. Additionally, θ_t, year dummy is included to account for all temporal variation in the region between 2004/05 and 2011/12. The standard errors are clustered at the village level, and robust errors were reported. Fixed effect regression was used to reduce the bias produced by a non-random assignment of households into off-farm employment. This model was chosen as this would clear away all time-invariant observable and unobservable household characteristics. Additionally, this takes care of any possible endogeneity due to unobserved time-invariant heterogeneity.

5.2 *Estimation of Heterogeneous Impact*

In the next step, we want to estimate the heterogeneous impact of off-farm income and labour allocation decisions of operator and spouse on food insecurity, using quantile regression. Quantile regression estimates treatment effects at different quantiles of the outcome distribution. This has an advantage over ordinary least squares, where the impact is estimated at the mean. In order to take advantage of panel data, we run a fixed effects panel quantile regression. Here the following model is considered:

$Y_{it} = U^*_{it}(1 + D_{it})$ where $U^*_{it} \sim U(0, 1)$ and a structural quantile function (SQF) for the variable of interest is estimated:

$$D_{it} = \alpha_t(\tau) + \beta_1(\tau)E_{it}, \quad \tau \in (0, 1)$$

The SQF defines the quantile of the latent outcome variable, D_{it}, which represents FIS—share of food expenditure in total expenditures, and ranges from 0 to 1. The quantiles were further classified into four classes, based on their magnitude of FIS as defined above (Smith and Subadoro 2007; COCA 2006; U.S. Department of Labor 2006). The estimated percentage change in the share of food expenditures, resulting in the change in FIS class, due to off-farm income and labour allocation is represented by β_l. This is estimated over time at the lth quantile of the share of food expenditures. α_{ht} is the fixed effect based on age of household head (HH) and spouse, years of education of both HH and spouse, household size, number of older persons ($\geq$64 years), number of children ($\leq$14 years), BPL cardholders (Yes/No) and wealth index.

We follow the procedure proposed by Powell (2013). According to him, the additive fixed effect framework used by Canay (2011) allows the parameters to vary based on the part of the disturbance term, while it excludes the other part assuming that it is fixed across time. But the motivation for the use of quantile treatment effects (QTEs) is to allow the parameters of interest to change based on the non-separable disturbance term U_{it}^* and not U_{it} which is assumed in additive fixed effects model. Moreover, in some cases when researchers are interested in estimating QTEs for the outcome variable, Y_{it}, they assume that they are not identified cross-sectionally. This method allows the parameters to be interpreted as in the cross-sectional quantile case.

6 Result and Discussion

Table 3 presents the parameter estimates from the fixed effects panel regression and panel quantile regression. In terms of sources of income and their impact on the food insecurity, results in Table 3 (column 2) reveal that the share of farm income (SFI) has a positive and significant impact on FIS, while off-farm income and business income did not have any significant effect on FIS. Household size has a positive impact on FIS while wealth has a significant negative effect on FIS. We should bear in mind that households with more members would tend to spend more on food items, and wealthier households would spend more on non-food items. Findings are consistent with theory. To control for overall variations over time, we included a time dummy (=1 if year 2011/12). The significant and negative effect of the time dummy variable indicates that, all other things being equal, FIS for rural households in India has fallen by about 6% between 2004/05 and 2011/12. This finding may suggest that rural households in India have become more food secure. Due to the limited variation over time, most of the estimated coefficients explaining household characteristics are not significant. Secondly, the results from the quantile fixed effects show the heterogeneous impact of off-farm income and labour allocation decisions of operator/spouse on FIS. Due to the space limitation, we only present selected

Table 3 Parameter estimates of determinants of food security: fixed effects panel regression and panel quantile regression, rural households, India

	Food insecurity classification with respect to the percentage of food expenditures (%) in total expenditures										
Variables	Average	*Lowest food insecure class* Percentage of food expenditure in total expenditures (<=50%)		*Medium food insecure class* Percentage of food expenditure in total expenditures (≥50% and <65%)			*High food insecure class* Percentage of food expenditure in total expenditures (≥65% and <75%)			*Highest food insecure class* Percentage of food expenditure in total expenditures (≥75%)	
		Selected Quantiles									
	Mean	0.10	0.15	0.20	0.30	0.40	0.50	0.60	0.70	0.80	0.90
HH age (years)	0.0001 (0.0005)	−0.000 (0.000)	−0.001*** (0.000)	−0.0014*** (0.0003)	−0.0011*** (0.0001)	−0.0008*** (0.0003)	0.0002 (0.0002)	0.0005 (0.0003)	−0.0000 (0.0002)	−0.0005*** (0.0002)	−0.0004* (0.0002)
Spouse age (years)	−0.0002 (0.0005)	0.0005 (0.0004)	0.0015*** (0.0003)	0.0018*** (0.0003)	0.0016*** (0.0001)	0.0009*** (0.0003)	−0.0001 (0.0003)	0.0000 (0.0002)	0.0004* (0.0003)	0.0008*** (0.0002)	0.0006*** (0.0002)
HH education (years)	−0.0005 (0.0005)	−0.0018*** (0.0005)	−0.0008 (0.0007)	−0.0003 (0.0003)	−0.0001 (0.0003)	−0.0007 (0.0005)	0.0007 (0.0005)	−0.0008*** (0.0003)	−0.0011*** (0.0002)	−0.0009*** (0.0003)	−0.0010*** (0.0003)
Spouse education (years)	0.0003 (0.0009)	−0.0030*** (0.0003)	−0.0038*** (0.0005)	−0.0043*** (0.0003)	−0.0043*** (0.0004)	−0.0036*** (0.0002)	−0.0031*** (0.0002)	−0.0029*** (0.0003)	−0.0026*** (0.0004)	−0.0026*** (0.0002)	−0.0020*** (0.0001)
Number of adults ≥64	−0.0024 (0.0033)	−0.0135*** (0.0024)	– 0.0136*** (0.0015)	−0.0139*** (0.0013)	−0.0091*** (0.0020)	−0.0074*** (0.0011)	−0.0075*** (0.0010)	−0.0034** (0.0016)	−0.0036*** (0.0007)	−0.0028** (0.0011)	−0.0031*** (0.0011)
Number of children in household ≤14	0.0027 (0.0017)	0.0054*** (0.0007)	0.0029*** (0.0005)	0.0046*** (0.0008)	0.0050*** (0.0008)	0.0026*** (0.0006)	0.0026*** (0.0004)	0.0051*** (0.0007)	0.0046*** (0.0004)	0.0060*** (0.0009)	0.0038*** (0.0004)
Household size	0.0024*** (0.0006)	0.0064*** (0.0005)	0.0064*** (0.0007)	0.0052*** (0.0005)	0.0046*** (0.0007)	0.0054*** (0.0002)	0.0043*** (0.0002)	0.0036*** (0.0005)	0.0027*** (0.0003)	0.0017*** (0.0005)	0.0008*** (0.0002)
Wealth index	−0.0422*** (0.0020)	−0.0472*** (0.0015)	−0.0536*** (0.0012)	−0.0506*** (0.0011)	−0.0500*** (0.0010)	−0.0513*** (0.0009)	−0.0478*** (0.0008)	−0.0449*** (0.0005)	−0.0445*** (0.0008)	−0.0379*** (0.0005)	−0.0334*** (0.0010)

(continued)

Table 3 (continued)

Variables	Food insecurity classification with respect to the percentage of food expenditures (%) in total expenditures										
	Average	*Lowest food insecure class* Percentage of food expenditure in total expenditures (<=50%)		*Medium food insecure class* Percentage of food expenditure in total expenditures (≥50% and <65%)			*High food insecure class* Percentage of food expenditure in total expenditures (≥65% and <75%)			*Highest food insecure class* Percentage of food expenditure in total expenditures (≥75%)	
		Selected Quantiles									
	Mean	0.10	0.15	0.20	0.30	0.40	0.50	0.60	0.70	0.80	0.90
BPL cardholders (yes/no)	−0.0028 (0.0027)	−0.0049** (0.0021)	−0.0037* (0.0021)	−0.0042** (0.0016)	−0.0034*** (0.0012)	−0.0085*** (0.0025)	−0.0078*** (0.0008)	−0.0049*** (0.0005)	−0.0064*** (0.0014)	−0.0049*** (0.0006)	−0.0050*** (0.0010)
Share of farm income (%)	0.0609*** (0.0155)	0.0952*** (0.0136)	0.0989*** (0.0166)	0.0676*** (0.0053)	0.0473*** (0.0052)	0.0420*** (0.0123)	0.0676*** (0.0131)	0.0356*** (0.0049)	0.0506*** (0.0057)	0.0459*** (0.0105)	0.0401*** (0.0059)
Share of business income (%)	−0.0002 (0.0092)	0.0216*** (0.0076)	0.0129 (0.0094)	−0.0032 (0.0053)	−0.0263*** (0.0029)	−0.0242*** (0.0077)	−0.0160*** (0.0049)	−0.0207*** (0.0040)	−0.0241*** (0.0034)	−0.0158*** (0.0046)	−0.0134*** (0.0035)
Share of off-farm income (%)	0.0039 (0.0079)	0.0098 (0.0092)	0.0112 (0.0092)	−0.0088* (0.0048)	−0.0170*** (0.0029)	−0.0212*** (0.0043)	−0.0063 (0.0063)	−0.0173*** (0.0046)	−0.0096** (0.0038)	−0.0111** (0.0046)	−0.0080*** (0.0026)
Only spouse permanent off-farm worker	0.0041 (0.0166)	−0.0194 (0.0153)	0.0151 (0.0261)	−0.0314** (0.0147)	0.0163 (0.0100)	0.0233*** (0.0086)	−0.0005 (0.0058)	0.0254*** (0.0051)	0.0142 (0.0103)	0.0309*** (0.0110)	0.0090 (0.0059)

(continued)

Table 3 (continued)

	Food insecurity classification with respect to the percentage of food expenditures (%) in total expenditures										
Variables	Average	*Lowest food insecure class* Percentage of food expenditure in total expenditures (<=50%)		*Medium food insecure class* Percentage of food expenditure in total expenditures (≥50% and <65%)			*High food insecure class* Percentage of food expenditure in total expenditures (≥65% and <75%)			*Highest food insecure class* Percentage of food expenditure in total expenditures (≥75%)	
		Selected Quantiles									
	Mean	0.10	0.15	0.20	0.30	0.40	0.50	0.60	0.70	0.80	0.90
Only operator permanent off-farm worker	−0.0087 (0.0069)	0.0108 (0.0094)	0.0124*** (0.0046)	0.0035 (0.0033)	0.0024 (0.0030)	0.0008 (0.0056)	−0.0086** (0.0034)	−0.0004 (0.0020)	−0.0198*** (0.0038)	−0.0029** (0.0014)	−0.0009 (0.0032)
Only spouse casual off-farm worker	−0.0092 (0.0063)	−0.0089 (0.0083)	−0.0178*** (0.0051)	−0.0105** (0.0047)	−0.0095*** (0.0034)	−0.0120** (0.0054)	−0.0286*** (0.0032)	−0.0183*** (0.0027)	−0.0067*** (0.0012)	−0.0029* (0.0017)	−0.0003 (0.0028)
Only operator casual off-farm worker	0.0030 (0.0038)	0.0160*** (0.0053)	0.0131*** (0.0026)	0.0131*** (0.0035)	0.0065** (0.0026)	0.0075 (0.0056)	0.0017 (0.0021)	0.0089*** (0.0018)	−0.0062** (0.0030)	0.0040** (0.0019)	−0.0007 (0.0024)
Both permanent off-farm workers	0.0004 (0.0167)	0.0022 (0.0083)	−0.0004 (0.0163)	−0.0301*** (0.0108)	−0.0332*** (0.0073)	−0.0126 (0.0127)	−0.0105 (0.0107)	0.0149** (0.0071)	−0.0001 (0.0045)	−0.0022 (0.0061)	−0.0267*** (0.0065)
Both casual off-farm workers	0.0065 (0.0042)	0.0070 (0.0044)	0.0037 (0.0045)	0.0092*** (0.0031)	0.0056** (0.0024)	0.0091* (0.0050)	0.0016 (0.0022)	0.0099*** (0.0014)	0.0002 (0.0028)	0.0146*** (0.0013)	0.0112*** (0.0012)

(continued)

Table 3 (continued)

	Food insecurity classification with respect to the percentage of food expenditures (%) in total expenditures										
Variables	Average	*Lowest food insecure class* Percentage of food expenditure in total expenditures (<=50%)		*Medium food insecure class* Percentage of food expenditure in total expenditures (≥50% and <65%)			*High food insecure class* Percentage of food expenditure in total expenditures (≥65% and <75%)			*Highest food insecure class* Percentage of food expenditure in total expenditures (≥75%)	
		Selected Quantiles									
	Mean	0.10	0.15	0.20	0.30	0.40	0.50	0.60	0.70	0.80	0.90
Operator permanent; spouse casual off-farm worker	−0.0027 (0.0147)	−0.0147* (0.0088)	−0.0080 (0.0147)	0.0207** (0.0090)	−0.0086** (0.0034)	0.0055 (0.0060)	0.0201*** (0.0039)	0.0218** (0.0092)	−0.0077 (0.0080)	−0.0168*** (0.0060)	−0.0166 (0.0125)
Spouse permanent; operator casual off-farm worker	−0.0041 (0.0155)	0.0563*** (0.0147)	0.0393*** (0.0142)	0.0208*** (0.0064)	0.0069 (0.0113)	0.0124 (0.0100)	0.0043 (0.0086)	0.0107 (0.0088)	0.0032 (0.0083)	−0.0100 (0.0096)	0.0196** (0.0085)
Year (=1 if year 2011–12)	−0.0619*** (0.0037)	−0.0572*** (0.0043)	−0.0579*** (0.0030)	−0.0659*** (0.0017)	−0.0623*** (0.0018)	−0.0634*** (0.0014)	−0.0677*** (0.0020)	−0.0638*** (0.0020)	−0.0730*** (0.0008)	−0.0687*** (0.0008)	−0.0697*** (0.0011)
Constant	0.6579*** (0.0207)										
Observations	32,268	32,268	32,268	32,268	32,268	32,268	32,268	32,268	32,268	32,268	32,268
Number of groups	17,139	17,139	17,139	17,139	17,139	17,139	17,139	17,139	17,139	17,139	17,139

Standard errors in parentheses; $^{***}p < 0.01$, $^{**}p < 0.05$, $^{*}p < 0.10$

quantiles across FIS classes; households with more than 65% of food expenditures in total expenditures are considered as highly FIS.

The age of the HH has a negative impact on FIS, while the spouse's age has a positive impact on FIS. These variables are significant at the 20th, 30th, 40th, 80th and 90th quantile. The educational attainment of the operator has a negative effect on FIS for 60th and higher quantiles. On the other hand, the educational level of spouses has a negative and significant effect on FIS for all quantiles. It is not only the direction of impact which is significant; the magnitude also reveals that spouse's education has a greater impact on reducing FIS (i.e. lowers the share of food expenditures in total spending) compared to operator's educational attainment. In other words, an additional year in spouse's education reduces FIS by 0.3% points (Table 3), across most quantiles. On the other hand, an additional year in operator's education decreases FIS by 0.1% points (Table 3) for 60th and higher quantiles. Findings here underscore the importance of educating women. This finding is consistent with that of Quisumbing et al. (1995). Policies designed to subsidise the education of girls and women can perhaps result in better jobs in the rural non-farm sector and could improve rural households' FIS status.

The number of older persons in a rural household has a negative and significant impact on FIS (Table 3) for all quantiles. On the other hand, the number of children in a rural household has a positive and significant impact on FIS (i.e. increase in food share expenditures), at all quantiles. A possible explanation is that compared to younger members, older people generally spend less on food items but more on health care. The coefficient on household size is positive and statistically significant for all quantile estimates, implying that larger families have higher food expenditures and hence more FIS. For example, an additional household member increases FIS in a household by about 0.1–0.6% points for all food expenditure quantiles, and the impact seems to be higher for the medium and high FIS classes (0.4–0.6% points).

Wealth has a significant negative impact on FIS. Specifically, for the high and highest FIS classes, columns 8–12 of Table 3 indicate that a one-unit increase in wealth index reduces FIS by about 3–4% points. The magnitude of the impact of wealth reveals that wealth could be a major factor in reducing FIS in rural Indian households. The share of BPL cardholders has a negative and significant effect on FIS and is consistent across all classes and quantiles. These findings are consistent with our expectations. It should be noted that BPL cardholders receive food entitlements from the government under various schemes, including the PDS and National Food Security Act (NFSA), and these programmes cover expenditure on staples such as rice, wheat, sugar and pulses. BPL cardholders are protected from any price shocks, which therefore make them less food insecure. All other things being equal, a unit increase in BPL membership would decrease FIS by 0.5–0.7% points for the highest FIS class (columns 11–12, Table 3).

We will now illustrate the effect of off-farm income and labour allocation decisions by operators and spouses, variables of interest to this paper. The share of farming income has a positive and significant impact on FIS. These findings suggest that farming income that is variable in nature may increase FIS. Surprisingly, the share of farming income has relatively less impact on the highest FIS class (columns 11–12,

Table 3) than on the least FIS class (columns 2–3, Table 3). This could be due to small landholdings owned by the highest FIS class. A unit increase in the share of farming income the FIS of the highest class by about 4% points; for the least FIS class, by about 9% points.

On the other hand, off-farm income has a negative and significant effect on FIS, implying that off-farm income reduces FIS for the medium class to the highest class. A unit increase in off-farm income share reduces FIS by 0.8–1.7% points, respectively, for the medium class to the highest class. This illustrates the importance of off-farm income in reducing the FIS of rural Indian households. This finding reiterates the contribution of rural non-farm income to stabilise household incomes and their consumption bundle. Perhaps off-farm income serves as a buffer in case of production and food shocks experienced by rural households. Similarly, business income has a negative and significant effect on FIS for the medium to highest classes. A unit increase in business income share reduces FIS by 1.3–2.5% points for the medium to the highest class. The above findings underscore the importance of income diversification in achieving FIS for rural Indian households.

Let us turn our attention to the labour allocation decision of operators and spouses, and its impact on FIS. A spouse's engagement in permanent off-farm work has a significant positive effect on FIS for the high and highest classes (the 40th, 60th and 80th quantiles). For other classes, the impact was heterogeneous. The lack of significance could also be explained by the low proportion of rural spouses engaged in permanent jobs off the farm. However, operators engaged in permanent off-farm work have significantly lower FIS. With respect to casual work, findings in the table show the opposite effect. For example, a spouse's engagement in casual off-farm work tends to reduce FIS and hence makes rural households less food insecure; the impact is significant across all classes or quantiles. This result was consistent with the findings of Quisumbing et al. (1995) and Thomas (1991). Specifically, a unit increase in spouse's involvement in casual off-farm work reduces FIS by 2.8–0.03% points. In the case of the operator, the impact on FIS was positive and significant for most of the classes.

Another interesting variable is operator and spouse engagement in casual off-farm work. In this case, casual off-farm work by both makes rural households more food insecure, especially in the medium and highest classes (columns 5–7 and 11–12, Table 3). For example, a unit increase in casual work by both operator and spouse increases FIS by 0.05–0.09% points for the medium class, and by 1.1–1.4% points for the highest FIS class. This is not surprising, because most of the poor households engaged in casual off-farm work and, given the low pay and poor working conditions for casual work, it would have a positive impact on FIS. Finally, the time dummy variable has a negative and significant coefficient, suggesting that over time (2004/05–2011/12), FIS has fallen consistently across all classes. Findings here shows that rural Indian households may be becoming more food secure. This result corroborates the estimates of the Planning Commission of India (2014).

7 Conclusion

India is mostly an agrarian economy, with almost 69% of Indians living in rural areas; 58% of rural households are directly dependent on agriculture for their employment and livelihood (NSSO 2104). According to the Planning Commission of India (2014), a significant number of rural households are still living below the poverty line and therefore may be food insecure. Furthermore, there have been increasing cases of malnutrition among children in rural households. Therefore, there is a need to assess and implement policies pertaining to enhancement of incomes and reduction in the food insecurity status of rural households. In this chapter, we evaluate the impact of off-farm income and labour allocation decisions, by operators and spouses, on the food security status of rural Indian households. We found that off-farm activities and off-farm business income reduce the food insecurity of rural households. But surprisingly, an increase in own-farm activities led to an increase in food insecurity. Therefore, promotion of off-farm activities and off-farm business, such as a sole proprietorship or small business, may help to enhance food security among rural Indian households.

We also analysed the impact on the food security of the casual and permanent off-farm work status of operators and spouses. We found that where spouses worked casually off the farm, their households were food secure, while where operators worked casually off the farm, their households were food insecure. However, if both operator and spouse worked casually off the farm, their households were food insecure. We also found evidence that a spouse's education plays a greater role, relative to the operator, in reducing food insecurity. Therefore, policymakers should promote education, especially targeting female members of the family. Our study also found that BPL cardholders were less likely to be food insecure, perhaps an indication that government policies are working with their intended targets. Therefore, broad and efficient implementation of policies such as the PDS and *Antyodaya Yojana* (subsidised food scheme) could be encouraged. In other words, expansion of these programmes of free provision of essential food items, to a significant proportion of the population living below the poverty line, would most likely reduce instances of food insecurity. Hence, targeting would enhance the effectiveness of these public policies.

References

Afridi, F., Mukhopadhyay, A., & Sahoo, S. (2015). Female labor force participation and child education in India: evidence from the National Rural Employment Guarantee Scheme. *IZA Journal of Labor & Development*, 1–27.

Barrett, C., Reardon, T., & Webb, P. (2001). Non-farm income diversification and household livelihood strategies in rural Africa: Concepts, dynamics and policy implications. *Food Policy*, 315–331.

Berg, E., Bhattacharya, S., Durgam, R., & Ramachandran, M. (2012). *Can rural public works affect agricultural wages? Evidence from India.* CSAE Working Paper.

Bhargava, A. (2014). The impact of India's rural employment guarantees on demand for agricultural technology. *IFPRI Discussion Paper*, 01381.

Canay, I. A. (2011). A simple approach to quantile regression for panel data models. *The Econometrics Journal, 14*(3), 368–386.

COCA (Canada's Office of Consumer Affairs). (2006). The consumer trends report. Canada's Office of Consumer Affairs, Ottawa, Ontario.

Ellis, F. (2000). The determinants of rural livelihood diversification in developing countries. *Journal of Agricultural Economics*, 289–302.

Goodwin, B. K., & Mishra, A. (2004). Farming efficiency and the determinants of multiple jobs holding by farm operators. *American Journal of Agricultural Economics*, 722–729.

Kareemulla, K., Kumar, S., Reddy, K., Rao, C., & Venkateswarlu, B. (2010). The impact of NREGS on rural livelihoods and agricultural capital formation. *Indian Journal of Agricultural Economics*, 524–539.

Kimhi, A. (1994). Quasi maximum likelihood estimation of multivariate probit models: Estimating farm couples' labor participation. *American Journal of Agricultural Economics, 76,* 828–835.

Lanjouw, J., & Lanjouw, P. (2001). The rural non-farm sector: Issues and evidence from developing countries. *Agricultural Economics*, 1–23.

Lanjouw, P., & Murgai, R. (2009). Poverty decline, agricultural wages and non-farm employment in rural India: 1983–2004. *Agricultural Economics*, 243–263.

Lanjouw, P., & Shariff, A. (2004). *Rural nonfarm employment in India: Access, incomes, and poverty impact.* NCAER.

Mishra, A., & Goodwin, B. K. (1997). Farm income variability and supply of off-farm labor. *American Journal of Agricultural Economics*, 880–887.

National Sample Survey Office (NSSO). (2104). Key indicators of situation agricultural households in India (January–December 2013), Vol. 33 of 70th Round. New Delhi: National Sample Survey Office, Ministry of Statistics and Programme Implementation, Government of India, New Delhi, India.

Owusu, V., Abdulai, A., & Abdul-Rahman, S. (2011). Non-farm work and food security among farm households in northern Ghana. *Food Policy*, 108–118.

Pal, S. (1996). Casual and regular contracts: Workers' self-selection in the rural labor markets in India. *The Journal of Development Studies*, 99–116.

Planning Commission. (2014). Report of the expert group to review the methodology for measurement of poverty. The government of Indian, New Delhi.

Powell, D. (2013). A new framework for estimation of quantile treatment effects. RAND Working Paper, WR-824-1.

Quisumbing, A., Brown, L., Feldstein, H., Haddad, L., & Pena, C. (1995). *Women: The key to food security*. Washinton, D.C: Food Policy Report, The International Food Policy Research Institute.

Ruben, R., & Berg, M. V. (2001). Non-farm employment and poverty alleviation of rural farm households in Honduras. *World Development*, 549–560.

Smith, L., & Subandoro, A. (2007). Measuring food security using household expenditure surveys, Food security in practice, IFPRI, p. 147.

Thomas, D. (1991). Gender differences in household resource allocation. *Population and Human Resources Department. Living Standards Measurement Study*.

U.S. Department of Labor. (2006). Consumer expenditures in 2004. U.S. Bureau of Labor Statistics Report 992, Washington, D.C.

Yang, D. T. (1997). Education and off-farm work. *Economic development and cultural change*, 613–632.

4

Global Change and Investments in Smallholder Irrigation for Food and Nutrition Security in Sub-Saharan Africa

Munir A. Hanjra and Timothy O. Williams

Abstract Investments in irrigation contribute to poverty reduction and enhance food security. This paper considers irrigation investments more broadly in the context of **rural–urban linkages** and thus examines rural irrigation schemes and peri-urban and urban agriculture using freshwater, groundwater and wastewater. We present case studies from East, West and Southern Africa, while focusing on the imperative of smallholders and of food security and nutrition. Evidence from Big Data and telecoupling show that, amid global change and sustainability issues, irrigation development strengthens connections between humans and nature with notable benefits to food security. **Transforming investments** to feed the future generation require priority investments in irrigation, solar energy for groundwater pumping, groundwater development policy, and integration of peri-urban and urban agriculture into food systems. Equally important will be **no-regret interventions** in wastewater reuse, water storage and groundwater buffer, micro-irrigation, and wholesale reconfiguration of farming systems, through anticipatory investments, to safeguard food security and sustainability into the distant future.

1 Introduction

Food security and agriculture are top development priorities across Sub-Saharan Africa (SSA). The Sustainable Development Goals (SDGs) aim to end poverty, end hunger, enhance food security and nutrition, and double agricultural production for smallholders. The EU Agenda for Change and the African Union prioritise sustainable agriculture and smallholder irrigation as a strategy for poverty reduction in Africa. Agricultural transformation and the doubling of irrigated area is a key pillar

We are indebted to Munir Hanjra who contributed to this book prior to his death in April 2019.

M. A. Hanjra (Deceased)
International Water Management Institute (IWMI), Pretoria, South Africa

T. O. Williams (✉)
International Water Management Institute (IWMI), Accra, Ghana
e-mail: wiltmt@aol.com

of the Comprehensive Africa Agricultural Development Programme (CAADP) and Feed Africa Strategy (AfDB 2016). Irrigation is in the manifesto of many political parties across the Southern African Development Community (SADC). National policies support new irrigation investments in agricultural greenbelts and growth corridors in Malawi and Tanzania. World Bank (2008) and other donors have renewed calls for re-engaging in agricultural water management solutions to achieve water security for sustainable development (Grey and Sadoff 2007; Sadoff et al. 2015).

Public policies and investments are needed today to ensure sustainable food security—defined as sustained poverty reduction and access to nutritious and healthy food for all (Hanjra et al. 2017c). Investments in sustainable intensification of agriculture are required for human prosperity and global sustainability (Conceição et al. 2016; Rockström et al. 2016; Hanjra et al. 2017e). Investments in irrigation can contribute to poverty reduction and enhance food security through several impact pathways, including higher crop yield, higher food production, higher income and consumption, gains in employment, higher wage earnings, women's empowerment through female employment, lower food prices, year-round food availability, augmentation of household assets, public infrastructure, education and health, and greater human prosperity (Hussain and Hanjra 2004; Hanjra and Gichuki 2008). Irrigation also improves crop diversification and market access, which together have positive effects on dietary diversity. Investment in irrigation improves food security and reduces poverty—there is robust evidence from Asia (Hussain and Hanjra 2003, 2004; Shinkai et al. 2007; Fan et al. 2008; Pingali 2012; Ward et al. 2013; Unver et al. 2016; Giordano et al. 2017) and strong emerging evidence from SSA (Hanjra et al. 2009a, b; Burney and Naylor 2012; Wichelns 2014; Williams 2015). Smallholder irrigation also improves nutrition outcomes—there is inconclusive but new evidence from Africa (Burney et al. 2010; Dillon 2011; Alaofè et al. 2016; Hanjra et al. 2017c). Also, the following nutrition impact pathways are supported by strong evidence: (i) food production, (ii) income (agricultural and non-agricultural) and (iii) female employment.

This chapter focuses on smallholder agriculture and considers irrigation more broadly within the rural–urban continuum in three sub-sectors: **rural irrigation schemes** using surface and groundwater, **peri-urban** and **urban agriculture** using wastewater and groundwater for irrigation. We aim to illustrate the contribution of smallholder-irrigated agriculture to various dimensions of food security and the overall contribution of small farms to economic growth, within the bigger picture of structural transformation and global change. In-depth case studies into public investments in irrigation in South Africa, public–private partnerships in Zimbabwe, and sustainable intensification in Tanzania are presented, along with a broader review of evidence from East, West and Southern Africa, including Big Data and telecoupling. The analysis shows what irrigation contributes to the well-being of smallholders ('beneficiaries'), what smallholder farmers in irrigation contribute towards food security ('agents of change'), and what policy opportunities exist to enhance food security and target nutrition. Programmatic-level support and policy interventions are needed to shape future investments in irrigation and promote related investments in appropriate farming and agrifood systems to deliver sustainable transformations, the EU Agenda for Change and the SDGs across SSA.

2 Global Change and Transformation of Smallholder Agriculture in Africa

Global food production more than doubled during the past 50 years and outpaced population growth and food demand in all regions except SSA. Global population stabilisation seems unlikely this century and world population is projected to increase from the current 7.2 to 9.6 billion in 2050 and 10.9 billion in 2100 (Gerland et al. 2014). Much of the population growth is expected to occur in Africa, where the population is projected to increase from roughly 1 billion today to between 3.1 and 5.7 billion by the end of this century. That would make Africa's population density roughly equal to that of China today, with huge policy implications for future food security and related areas—environmental (natural resources, water quality, wastewater reuse), economic (jobs, wages, poverty, inequality, rural-urban divide, migration), social (crime, unrest), health (higher maternal and child mortality), government (investments in agriculture, irrigation, health, education, energy, infrastructure, water and sanitation) and climate change (biofuels, solar power, food miles, carbon emissions).

Africa has made tremendous gains in food security during the past decades. While the outlook for 2050 and beyond is encouraging, population growth, urbanisation, natural resource distribution patterns, climate change and, above all, future changes in diets (Keats and Wiggins 2017) pose complex challenges in terms of sustainably feeding the future generation and achieving the SDGs. Urbanisation drives up the demand for food products, such as premium rice, that are not supplied by local farmers, suggesting that net negative trade will increase without transformation. However, Africa has enough labour, land, water, energy and natural resources, and African food security prospects are brighter than ever (FAO 2015). Nonetheless, investments are needed to transform farming systems across Africa (Williams 2015; Dixon et al. 2017).

Three basic transformations are needed to ensure food security: structural, agricultural and dietary transformation (Timmer 2017). Structural transformation involves four main drivers including: a declining share of agriculture in national income and employment, a rising share of urban economic activity in manufacturing and modern services, migration of rural workers to urban areas, and demographic transition towards lower birth and death rates and better health standards. Indeed, structural transformation has been the main pathway towards food security and out of poverty for many of today's developed societies; it depends on rising productivity in both agricultural and non-agricultural sectors accompanied by a declining share of agriculture in income and employment (Timmer 2017). Data from 29 developing countries confirm that structural transformation raises total income and poverty falls faster with government support for smallholder agriculture, which in turn improves nutrition in rural areas (Webb and Block 2012).

Agricultural transformation is driven by changing demand for food in domestic markets, opportunities for food exports, commercialisation, intensification and diversification of agriculture and adoption of innovations in commodity value chains

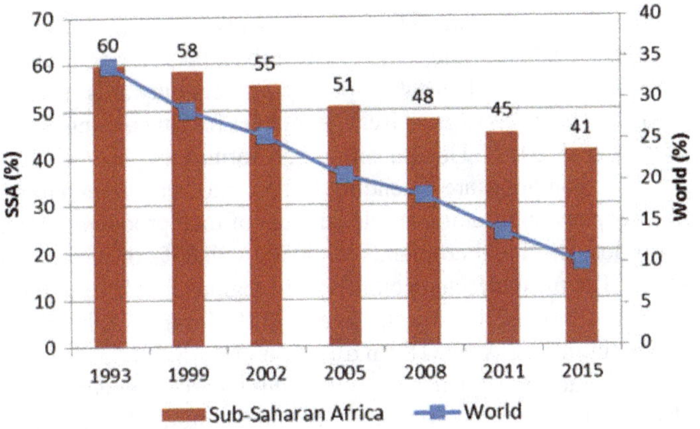

Fig. 1 Progress towards poverty reduction (below USD 1.90/day). *Data Source* World Bank

which together serve to raise productivity per ha and productivity per worker (Barrett et al. 2017; Timmer 2017). Africa shows early signs of "coupled growth", i.e. agricultural growth and structural transformation (e.g. in Ethiopia, Rwanda) (Abro et al. 2014; Barrett et al. 2017; Sheahan and Barrett 2017). Examples of country specific transformation include tenure security and irrigation investments in Tanzania, improved rice yields in Senegal and Mali, higher cotton yield in Burkina, scaling-up of agricultural innovations in Nigeria, warehousing receipt system in Uganda, land use consolidation policy and registration system in Rwanda, floriculture exports in Ethiopia and horticulture exports in Kenya (Verdier-Chouchane and Boly 2017).

But the challenge is in bringing to scale existing and successful interventions to enhance food security and reduce poverty for millions of households across Africa (Appendix). Poverty has declined overtime in SSA (World Bank 2018) (Fig. 1), and the inclusiveness of child health improvements has increased (Sahn and Younger 2017). The number of undernourished increased from 177 million (2005) to 232 million (2017), with rapid growth in almost all sub-regions in recent years, especially in Western Africa (FAO 2019). This situation is even worse in conflict-affected countries of SSA (where undernourished people increased by 23.4 million between 2015 and 2018) and in drought-sensitive countries (where undernourished people increased by 45.6% since 2012). Thus, major investments are needed in agriculture itself to support structural transformation and governments must provide strong support to ensure food security for their citizens across Africa.

2.1 *Global Change and Rural–Urban Linkages*

Rapid population and income growth and greater rural–urban linkages are increasing demand for energy, food and water across Africa. Supportive developments have also taken place in regional trade liberalisation and food markets, strengthening institutions and policies, social media, and investments in human capital and modern technology. Amid such drivers and trends, rural–urban linkages and farming systems are changing dynamically. Urbanisation pushes farmers outwards, such that many begin to farm on city outskirts to grow fresh vegetables for cities, using wastewater for irrigation (Hanjra et al. 2012). This transition transforms the role of peri-urban areas in food security. Urban areas are extending further into peri-urban and rural areas, such that urban expansion is taking arable land out of food production—resulting in about 1.8–2.4% loss of global croplands by 2030, with 85% of this taking place in Asia and Africa. The most affected countries and regions in Africa in terms of crop production loss are Egypt (36% loss), Nigeria (12% loss), and the region around Lake Victoria basin in East Africa (Bren d'Amour et al. 2016).

Urbanisation impacts the dynamics of city region food systems and the food mix demanded by urban consumers—reflecting income, cultural diversity and lifestyle (westernisation of diets—Pingali 2007)—and this, in turn, influences food production and global food supply chains (Schmidt et al. 2015). Indeed, dynamic areas around modern cities are subject to rising population pressure, with cities contributing to the uncoupling of food consumption and local agriculture and reducing capacity for food self-sufficiency (Tedesco et al. 2017). There is ample opportunity to recouple food production and consumption in urban and peri-urban agriculture through recovery and reuse of nutrients, wastewater and energy from urban sanitation systems in agriculture, but investments are needed to accomplish this (Otoo and Drechsel 2017).

2.2 *The Imperative of Smallholders and Food Security*

There are about 475 million farms in the developing world (Rapsomanikis 2015). They contribute to food security, sustainable use of natural resources, rural economies and livelihoods. Recent assessments suggest that growth in smallholder agriculture can have strong impacts on poverty reduction and food security (IFAD 2011; Lowder et al. 2016). Also, the success of rural development strategies in Africa depends on small farms (Salami et al. 2010; Wiggins et al. 2010; Larson et al. 2016).

Smallholders make a notable contribution towards food security and global food production, producing nearly half of the world's food. Smallholder-dominated systems (<5 ha per household) in Latin America, SSA, and South and East Asia account for about 380 million farming households, 30% of agricultural land, 70% of food calorie production in these regions and more than half globally. They provide 70% of calories consumed compared with 50% globally (Leah et al. 2016). SSA has 43.55 million smallholder families (<2 ha), accounting for 80% of farms; 69% of the calories produced in smallholder systems, and 84% on farms in urban areas.

Smallholder food production within and around cities has the potential to enhance access to healthy and nutritious food; improve environmental quality by reducing urban heat island effect, emissions and storm water runoff; promote local circular economies; create new jobs in peri-urban areas; reduce dependence on expensive food imports; and support agricultural transformation that reduces poverty (Zezza and Tasciotti 2010; Grewal and Grewal 2012). Achieving the imperative of local food security through self-reliance requires effective use of local land and water resources and investment and policy support from city authorities and governments to promote safe reuse of wastewater and nutrients in urban and peri-urban food production systems. This will support a circular metabolism model for enhanced urban resilience (Hanjra et al. 2017d).

> We need the World Bank, we need the IMF, we need all the big foundations, we need all the governments to admit that for 30 years we all blew it, including me, when I was President. We blew it. We were wrong to believe that food is like some other product in international trade. And we all have to go back to a more environmentally responsible, sustainable form of agriculture. We should go back to a policy of maximum **food self-sufficiency**.
>
> Former US President Bill Clinton, 2008.

3 Methodology and Conceptual Framework

The overall objective of this paper is to showcase what works where, and agricultural transformation pathways to enhance future food security and prosperity in Africa through renewed investments in smallholder-irrigated farming systems and supportive public policies. This is in contrast to the usual portrayal of irrigation success or failure and deep-rooted factors limiting paths to successful transformation in Africa. Our focus is on steering local change towards a brighter future outlook for shared prosperity for the people of Africa. To end poverty (SDG 1), end hunger and enhance food security and nutrition (SDG 2) in Africa, our conceptual framework no longer focuses on conventional irrigation using freshwater only, but instead presents wider options—to ensure sustainable water management and sanitation (SDG 6), sustainable energy (SDG 7) using solar power for groundwater irrigation, and sustainable production and consumption patterns (SDG 12). Thus, another major methodological departure is the broader focus on irrigation along the rural–urban continuum to link directly to realities on the ground across Africa. We chose case study data and methods to examine smallholder rural irrigation schemes and peri-urban and urban agriculture systems using surface water, groundwater, and wastewater. Rural–urban linkages and solar energy options help to provide more promising perspectives on sustainable water management solutions for enhancing food security, within the urban sanitation-agriculture interface and green energy solutions.

We present findings from studies across East, West and Southern Africa, using mixed methods. National irrigation scheme (NIS) databases for South Africa, Tanzania and Zimbabwe were used to select several irrigation schemes in each country for in-depth case studies, involving extensive field data collection missions, face-to-face interviews with irrigators and other stakeholders, workshops with irrigation authorities, panel discussions with local leadership and district authorities, regional water policy dialogue, and presentations to donors and global experts[1] (Riesgo et al. 2016). We applied extensive data filtering to select and review recent studies published in peer-reviewed journals to illustrate the contribution of irrigation investments to food security in SSA, including the Middle East and North Africa (MENA). Our primary focus was on food and nutrition security, within the broader context of the SDGs and predicted climatic and demographic changes.

4 Results

There is widespread consensus that past investments in irrigation have enhanced food security, alleviated poverty and transformed agriculture across Asia (Fan et al. 2000, 2008; Hussain and Hanjra 2003, 2004; Kurosaki 2003; Pingali 2015; Mishra et al. 2017). However, relatively little has been published about the poverty and food security impacts of past investments in irrigation in SSA, though there is strong evidence from new studies on smallholder irrigation across Africa (Hanjra and Gichuki 2008; Hanjra et al. 2009a, b; Burney and Naylor 2012; Hagos et al. 2017; Woodhouse et al. 2017). Evidence on nutrition outcomes is, however, relatively limited (Burney et al. 2010; Dillon 2011; Alaofè et al. 2016; Hanjra et al. 2017c).

Studies from **Nigeria, Mozambique, Tanzania, Uganda** and Nepal show that agricultural production has direct and important linkages with household dietary and nutrition outcomes (Carletto et al. 2015), while data from Indonesia, **Kenya, Ethiopia, and Malawi** show that empirical evidence on the link between production and consumption diversity is weak (Sibhatu et al. 2015). Thus, the evidence on agriculture and nutrition outcomes remains inconclusive. The lack of evidence on the impact of agricultural programmes on nutrition outcomes is interpreted as a reflection of the weakness in programme design and implementation, lack of rigour in evaluation, and more importantly the fact that emphasis on irrigation/agriculture and nutrition outcomes is relatively new (Carletto et al. 2015; Pingali 2015). However, for the first time, SDG 2 has a dedicated target to 'end all forms of malnutrition'. Programmatic support for agriculture and irrigation interventions can play a direct role in enhancing food security and nutrition, but nutrition outcomes may depend on local conditions and the state of the economy (Domènech 2015; Fiorella et al. 2016; Pandey et al. 2016). Evidence-based interventions and nutrition-sensitive approaches are needed to support agricultural programmes to achieve SDG 2 targets on food and nutrition security (Hanjra et al. 2017c).

[1]This work was presented at the European Commission Joint Research Centre workshop held in 2015 in Seville, Spain (see Acknowledgements).

4.1 *Rural Irrigation Schemes Using Surface and Groundwater*

4.1.1 Public Investments: South Africa

South Africa's national irrigation development plan prioritised the rehabilitation of old irrigation schemes and the development of new schemes to benefit smallholders. Research evidence has shown that irrigation has a very high social value for subsistence farmers and public investments offer remarkable returns in food security and social equality. The NIS database shows that South Africa has about 1.3 million ha under irrigation, but only 47,670 ha in 302 schemes are under smallholder irrigation (van Averbeke et al. 2011). Located in former homelands, these schemes have the potential to contribute to food security and income and to create employment for the poor.

IWMI's Southern Africa regional office based in Pretoria, South Africa (IWMI SA) and national partners have accumulated vast knowledge over the past 20 years on agronomy, water management and revitalisation of irrigation schemes in Southern Africa. However, knowledge gaps remain in water economics, financial and investment analysis and policy measures to address the needs of smallholders and poor farmers. Smallholder irrigators lack experience, motivation, funds, assets, machinery and marketing skills to take full advantage of government funding targeting the rehabilitation of irrigation schemes to increase productivity, profitability and social integration of new entrants into the national fabric.

Data from irrigated areas in the **Limpopo basin** show that rural poverty varies widely between the basin states: in Zimbabwe (69%), Mozambique (68%), South Africa (56%) and Botswana (20%). Food insecurity had a lower range (0–40%) than poverty (0–95%). 'Areas of high food insecurity and poverty consistently coincide with areas experiencing low water availability' (Magombeyi et al. 2016: 20). Indeed, households with access to irrigation in the Limpopo province of South Africa are more food secure (86%) compared to dryland farmers (53%) (Oni et al. 2011). Irrigation increases crop yield, income, assets and farm diversification to enhance household food security. To further enhance food security, irrigators should be targeted for education and extension services, while food aid should be targeted to those households without access to irrigation and those with large families and few assets (e.g. agricultural land and tools). Policymakers need to support irrigation to assist farmers to produce their own food, to help break the food aid 'dependency syndrome'. Nutrition should be part of the public agriculture programme to promote a paradigm shift in local eating habits to combat undernutrition and obesity.

Despite public funding, amid poverty and lack of skills, smallholder farmers continue to prioritise low value crops for food security reasons. Data from the Mooi river irrigation scheme of **KwaZulu-Natal**, South Africa show that farmers applied less water (62%) to their potato crop (with a water value of USD 0.25 per m^3) than to crops with a lower water value, such as maize (USD 0.12 per m^3) and beans (USD 0.10 per m^3) that are important for household food security (Muchara et al.

2016). As water is provided free to farmer groups, unequal distribution at tails and inefficient water use are common in this scheme. This suggests that cost recovery mechanisms and user participation policies can incentivise efficient water use. This observation is supported by evidence from the Tugela Ferry irrigation scheme in the Msinga District (Fanadzo 2012; Sinyolo et al. 2014; Maziya et al. 2017).

Data from 223 small farming households in the **Eastern Cape**, one of the poorest provinces in South Africa, show that maize yield will be positively affected by climate change under rainfed or irrigated conditions, while potato yield will decline. Both institutional and infrastructural support, through access to credit and irrigation facilities, were recommended for adequate adaptation to future climate change impacts on food security (Hosu et al. 2016). Public policies and planning processes should carefully consider such yield trade-offs in making integrated policies to enhance food security and sustainability at scale.

4.1.2 Public Private Partnerships Model: Zimbabwe

The NIS database for Zimbabwe shows that the total area equipped for irrigation is 186,000 ha, with 130,000 ha currently functional and 56,000 ha in need of rehabilitation. Zimbabwe has a total potential irrigable area of 2.5 million ha that could be developed at a total cost of about USD 10 billion, for which government needs to attract private sector investment into new irrigation developments (Table 1) (Hanjra et al. 2016b). This area spans old resettlement, communal, agriculture authority and new settlement farming sectors. It could be irrigated using water from existing and planned dams, and small, medium and large rivers and groundwater. However, the main challenges remain investment and sustainability.

Table 1 Current status and investment opportunities in irrigated farming systems in Zimbabwe

Sub-sector	Equipped	Functional	Under Rehabilitation
Irrigated farming systems (ha, 2015)			
Communal	15,000	10,000	5000
Old resettlement (A1)	30,000	23,000	7000
New settlement (A2)	61,000	22,000	39,000
Agriculture development authority (ARDA)	17,000	12,000	5000
Plantations	63,000	63,000	0
National total	186,000	130,000	56,000
Investment opportunities			
	Equipped	New irrigation	Under Rehabilitation
Capital cost (estimated) (USD million)	–	10,000	196
Financing model	Public	Hybrid	Public–private

Source Department of Irrigation (Hanjra et al. 2016b)

Detailed fieldwork in 14 smallholder irrigation schemes in the **Masvingo province of Zimbabwe** showed that common **drivers of success** of enhanced food security are proximity to markets, access to credit, contract farming, farmer organisation, crop choices, extension support, infrastructure maintenance and storage facilities (Hanjra et al. 2016b). Output marketing is a serious challenge in almost all rural schemes, and there is a need for grading and standardisation of weights and measures and quality assurance. Indeed, even town and city market traders rarely use scales, but instead use buckets, and other indigenous measures. Irrigation schemes directly selling to local institutions such as residential colleges, schools, hospitals, churches, and mining workers tend to be more productive and profitable. Irrigators follow cropping plans at scheme level and acquire inputs at competitive prices through group bargaining. Smallholder irrigators here are generally much better organised than smallholders across the border in South Africa. Most irrigation schemes have clear by-laws and rules but no recourse beyond their associations. For instance, after the rehabilitation of some surface irrigation schemes, food production improved dramatically and smallholders entered into joint production and marketing contracts with some supermarkets and fast food giants in South Africa. However, in order to take full advantage of these opportunities they needed government support and legal assistance in contract negotiation and enforcement.

Value addition activities such as tomato paste, dried packaged beans, maize cereals, roasted peanuts, and premium grade organic food production can add further value, but require storage, processing facilities and linkages with distant urban markets. Mobile phones allow smallholders to enquire about prices and do business with supermarkets and bulk buyers, but poor transport facilities make it expensive to sell small quantities in town and city markets (Hanjra et al. 2016b). Farmer training in business management and financial accounting offers high returns on investment. Energy shortages are a serious problem, particularly where pumping is required to pressurise mini sprinklers. Reverting to **gravity irrigation** or using **solar panels** are potential solutions.

4.1.3 Sustainable Intensification Business Model: Tanzania

The National Irrigation Master Plan 2002 identified the total potential irrigation development area at 29.4 million ha, including 2.3 million ha of high potential area. However, the actual area under irrigation is only 450,962 ha (Tanzania NIS database 2015) and there are 2427 small-scale irrigation schemes (Hanjra et al. 2016a). Tanzania's National Water Policy and Water Resources Management Act 2009 provides strong support for irrigation and sustainable intensification development. In many cases, smallholders initiate proposals for irrigation scheme development and complementary rural infrastructure including schools, roads and health centres, and manage the scheme under local by-laws with guidance from irrigation authorities. In-depth studies in seven smallholder irrigation schemes in the Ruvu River Basin and Morogoro region show successful transition from crop diversification to sustainable intensification. This covers the whole range of crop production, but the major irrigated crop in Tanzania is rice. For example in the Ruvu Basin, rice yield has increased dramatically from 2 tonnes per hectare (t/ha) to about 5 t/ha (Fig. 2) and up to 8 t/ha with sustainable rice intensification (SRI). **Ruvu River branded rice is a successful business model**. Smallholders add further value through increased investment in output marketing, such as branded rice, smaller packaging and direct sale to town markets in partnerships with agribusiness. They also support infrastructure development such as local roads, storage, energy, water, schools and job training for irrigators to support transition to non-farm jobs in nearby towns and to attract private sector investment. Farmers are organised into Irrigator Associations and regular meetings ensure inclusivity and sustainability.

Irrigation schemes in the **Morogoro region** are a model for sustainable intensification. Here maize, paddy and beans are the predominant crops, but the sustainable intensification trajectory leads to the integration of vegetables first, followed by livestock and fruit orchards and finally fish in the most innovative schemes. For example, livestock integration into irrigated crop production in Kilosa district has enhanced food security and directly improved nutrition outcomes, due to greater milk and protein consumption at household level. Influx of large livestock herds during the dry season can cause serious damage to irrigation infrastructure. The **Kilosa district livestock grazing fee model is a business innovation** for enhancing profitability and sustainability. Some irrigation schemes have responded to the livestock challenge by offering fallow croplands to communal pastoralists/seasonal livestock herders to graze their animals on crop residues and after-harvest regrowth, in exchange for a levy per livestock unit (Hanjra et al. 2016a). The herder's livestock productivity improves (higher cow milk production and reduced calf mortality in the dry season) due to better quality and palatable feed. This business model brings new income for irrigators, which is partly used to maintain watercourses, while direct grazing or tethering brings free livestock manure to fallow fields to enhance soil fertility.

Many schemes practice irrigation using an irrigation business model, where smallholders pool their land and water resources to produce high value crops (chillies,

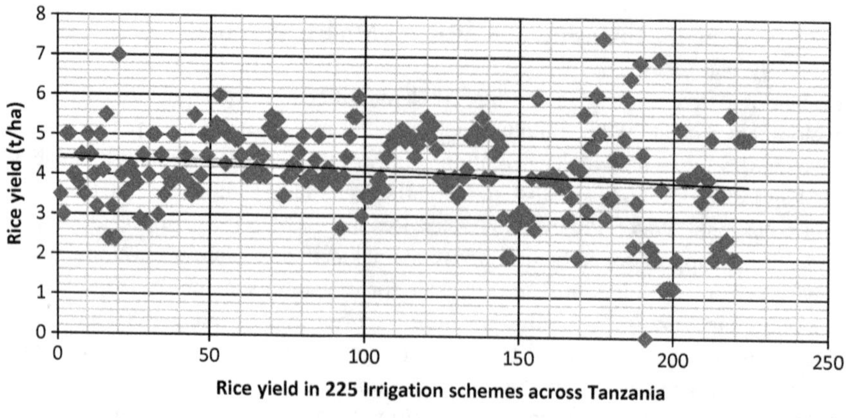

Fig. 2 Closing the yield gap, enhancing prosperity and sustainability (Hanjra et al. 2016a)

tomatoes and table grapes) for market sale through contracts with supermarket chains. Some schemes even undertake their own processing and marketing to sell branded rice (e.g. surface water irrigation, Ruvu basin), table grapes and bottled grape juice (e.g. groundwater irrigation, high-tech facility in Chamwino district). Indeed, the **Chamwino district grape production system is a business model innovation by smallholders**. Investment comes from commercial loans to install modern irrigation systems, such as pressurised drip irrigation using groundwater. Local political leadership, including district development authorities, provides strong support and loan guarantees to the agriculture development bank on behalf of smallholder farmers, who still work on their commercial farm as labourers, to hire service providers for technical works. Annual benefits are distributed equally among all the farmers (Hanjra et al. 2016a). Equal landholding and benefit sharing on business principles help to avoid equity problems. Member irrigators earn four times higher profit than non-irrigators, and many more farmers are registered to join the scheme on a first-come, first-served basis. This business model is gaining momentum and being scaled out to other irrigation districts.

4.1.4 West and East Africa

Past irrigation policy in many West African countries encouraged investment in large-scale irrigation for agricultural development and food security reasons, yet evidence of success remains inconclusive (Williams et al. 2012). Large-scale irrigation projects are expensive, per hectare and per person lifted out of poverty. Rice grown in some of these irrigation schemes could not compete until recently against cheaper imported rice in urban markets and was considered less attractive by large households and women shoppers (Demont et al. 2017). Due to low quality of local rice and high cost of production, rice production has low profitability, e.g. in **Niger** (Katic et al. 2013), **Benin** (Nonvide et al. 2017) and **Office du Niger**, Mali (USD 138 per ha) (Sidibé and Williams 2016). Farmers persist with irrigation as long as irrigation infrastructure works with minimum maintenance, as the real cost of rice production (4 tonnes paddy/ha) is high and and sometimes exceeds return (Comas et al. 2012). Irrigators therefore tend to maintain a diverse portfolio of livelihood activities, including rainfed agriculture and non-farm activities. Households combine irrigated rice with traditional rainfed and flood-recession crops to enhance agricultural incomes. Rice yield has improved remarkably, for instance, by 60% in Benin (Nonvide et al. 2017). However, the high cost of using irrigation due to irrigation water not being available all the time, means that farmers tend to move in and out of irrigated farming depending upon the availability of loans and investment needs. Higher rice yield, water use efficiency, intensive use of irrigated land and greater emphasis on market-oriented production will translate into greater success. Investment in small-scale irrigation is supply-shifting, offering higher benefit to gender-equitable poverty reduction, but must be complemented by investment in demand-lifting interventions such as quality upgrading, branding and market promotion to achieve desired results.

Studies linking solar irrigation and food security linkages are limited in number (Burney et al. 2010; Alaofè et al. 2016). Data from Kalale district in northern **Benin** show that compared to manual irrigation, solar-powered drip irrigation greatly improves crop production diversity and dietary diversity (Alaofè et al. 2016). Women irrigators increase their production of vegetable (25%) and fruit (55%) and consumption threefold, thereby improving household food and nutrition security. In addition, the purchase and consumption of other food items, including sorghum, oil, rice and fish also increased. Many women used their additional income on food (60%), health (55%), utilities (40%) and education (25%). Thus, solar irrigation offers potential to enhance household nutritional status through direct food consumption and to increase income to improve access to health and education.

Similar impacts on food security and poverty reduction have been widely reported for smallholder irrigation in **Ethiopia** using surface water (Hanjra et al. 2009a), groundwater (Hagos and Mamo 2014), both surface and groundwater (Zeweld et al. 2015), spate irrigation (Hagos et al. 2017) and in other countries (Fig. 3).

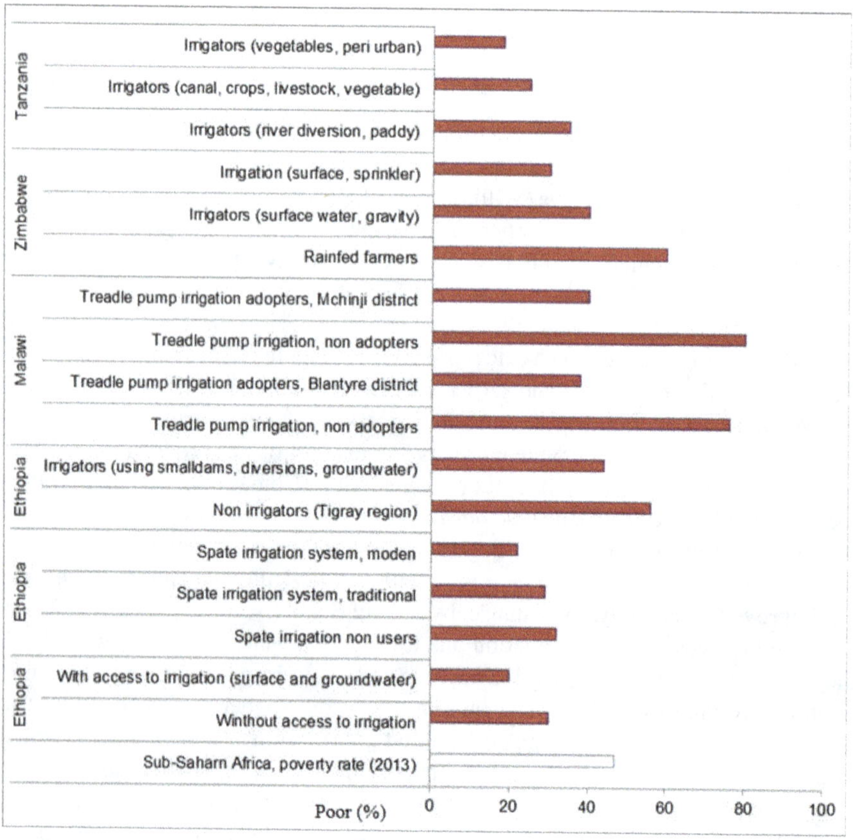

Fig. 3 Water management solutions and poverty reduction across SSA

Pathways linking irrigation with nutrition and health gains remain understudied. Only a few rigorous studies assess the linkages between irrigation and nutrition, but most show a positive effect of irrigation interventions on food security (Domènech 2015). In a review of (28) studies mainly focusing on SSA, only one study had assessment of nutrition outcomes as a primary objective. The study by Hagos et al. (2017), however, did not find any evidence of nutrition outcomes attributable to spate irrigation in Ethiopia. Other studies report mixed or inconclusive results. A study in Ghana (Namara et al. 2011) found inconclusive evidence on household dietary diversity score for rainfed versus groundwater irrigation. Data from **Burkina Faso** on household and child nutrition and dietary diversity measures showed an increase in household micronutrient-rich foods, such as dark green leafy vegetables and yellow or orange fruits, and maternal and child intake of leafy vegetables or eggs as a result of irrigation (Olney et al. 2015). A study in **northern Mali** (Dillon 2011) showed that between 1998 and 2006, households with access to irrigation greatly increased their daily calorie intake (1836 cal) compared to those without irrigation (925 cal), suggesting that irrigation helped to improve calorie intake over time. Also, a study in Zimbabwe that examined the linkages between irrigation and dietary diversity ranked independent irrigators (highest), and scheme irrigators, home gardens and non-irrigators (lowest), based on diversity of food produced and weekly food consumption (Moyo and Machethe 2016).

4.2 Urban and Peri-urban Agriculture Using Wastewater and Groundwater

Urban and peri-urban agriculture can contribute towards food and nutrition security and poverty reduction through more nutritient-rich food and direct market access than traditional irrigated agriculture producing cereals. The focus on fruits and vegetables supports, in particular, improved **nutritional benefits**. Various studies demonstrate the linkage between agricultural interventions and nutrition outcomes, showing that the production of targeted nutritient-rich crops, home gardens, and diversification of agricultural production systems towards fruits, vegetables and aquaculture can potentially improve nutrient intake and nutrition outcomes (Zezza and Tasciotti 2010; Pandey et al. 2016).

Peri-urban areas continue to expand fast amid urbanisation in Africa, creating the challenge of turning increasing quantities of wastewater and urban organic waste into opportunities for reuse and recycling in agriculture. With rising demand for fresh vegetables in cities, local production within and around urban areas across Africa is increasingly specialising in highly profitable irrigated vegetable production. For instance, **Accra, Ghana**, has some 800–1000 vegetable farmers cultivating unused open spaces near streams and drains within its core area, producing exotic vegetables (lettuce, cabbage, spring onions and cauliflower) and local vegetables (tomatoes, okra and chilli peppers). Sources of water are shallow wells and streams carrying wastewater. Watering cans are used for fetching wastewater from drains, which is very labour intensive as the hot climate especially in West Africa demands daily or twice-daily irrigation. Therefore, plots cultivated per farmer are usually small (0.01–0.05 ha). Motorised pumps allow larger plots and they are increasingly used and shared among farmers where the distance between water source and fields is long. But even then, farmers still use watering cans to draw water from on-farm storage reservoirs filled by pumping (Drechsel and Keraita 2014: 3).

High market demand, close market proximity and year-round availability of (waste) water are the main drivers of urban and peri-urban vegetable production (Drechsel and Keraita 2014). Vegetable farming is a profitable venture, such that two out of every three vegetable farmers were unwilling to leave even if they were offered regular salaried jobs. Potential health risks exist but have not stopped farmers from gaining a livelihood by using 'unsafe' water for irrigation as this leads to higher profits compared with rural rainfed farming on similar sized-plots. Other studies provide supportive evidence on the acceptance of wastewater use in urban and peri-urban areas in Morogoro, **Tanzania** (Samson et al. 2017), Bulawayo city in **Zimbabwe** (Makoni et al. 2016), Blantyre, Lilongwe and Mzuzu in **Malawi** (Msilimba and Wanda 2012; Holm et al. 2014), and Addis Ababa, **Ethiopia** (Weldesilassie et al. 2011). The main risk is less with the farmer, but with the large number of consumers who maybe unaware of the water source used. IWMI estimates that in Ghana, the 'beneficiaries' of urban vegetable production include about 2,000 urban farmers, 5,300 street food sellers and 800,000 daily consumers within the major cities (Drechsel and Keraita 2014: 3). Benefits from urban farming manifest in different ways as shown by reports across Africa summarized below.

In **Accra, Ghana**, the wastewater of about 225,000 residents—some 14% of the total urban population—currently has a 'natural' wastewater treatment system that is not disposal-oriented, but turns wastewater into an asset through its use in irrigated open-space farming. This number is probably larger than the one served by sewerage and existing treatment plants in the city (Lydecker and Drechsel 2010). Urban market gardening is generally practised on large open areas not used for other commercial purposes, or in home gardens (backyards). Overall, open-space farming mainly supports cash-crop niche markets (e.g. for exotic vegetables), while backyard farming supplies household food susbsistence needs and thus serve to reduce household food expenditure.

In Zambia, a quarter of urban households engage in urban agriculture, growing vegetables and other crops. Low-income city gardeners make USD 230 per year from sales (FAO 2012). Home gardening accounts for nearly half of fruit and vegetable production. In Lusaka, 90% of the residents practising urban agriculture in 2005 were women and for the majority it provided nearly one quarter to one half of their income, with 70% of growers cultivating small fields of less than 0.5 ha. A survey in four Zambian cities—Lusaka, Kabwe, Kitwe and Ndola—found that maize was the most frequently grown crop, but half of production consisted of horticultural crops: pumpkins, beans, onions, rape, tomatoes, groundnuts, sweet potatoes and cabbage (FAO 2012). Also, 80% of Lusaka's supply of leafy rape is produced locally and marketed through small vendors in city streets and neighbourhoods, while revenue from sales accounts for 18% of annual household income in Lusaka and about 50% in the other three cities (FAO 2012).

In Zimbabwe, around 70% of urban households practise some form of urban agriculture on residential land in the capital city of Harare, with about 17% practising it outside the residential properties in the vicinity of water bodies (Hanjra et al. 2017a). The streams and wetlands around the city often receive wastewater and this co-mingled water supports agricultural activities and contributes to food security and nutrition for the urban poor. Data around Bulawayo city for 2006 showed that some 500 farmers were using wastewater for irrigation of vegetable crops on small plots of half a hectare each (Mutengu et al. 2007). This is also supported by recent work (Makoni et al. 2016).

In Malawi, urban agriculture has social value for food security of poor households. About 25% of the population live in urban areas and use wastewater for irrigated agriculture (Msilimba and Wanda 2012). The 2030 Urban Structure Plan of Lilongwe City (GoM 2013) notes that 'urban agriculture mainly consists of illegal farming practised seasonally in open spaces in the city' and 'it should be regulated land use' for 'commercial farms in the future' (p. 26).

In East African capitals including **Addis Ababa**, **Dar-es-Salaam**, **Kampala** and **Nairobi**, the proportion of urban households that are farming (25–55%) has not diminished with urban growth, and urban farming households are better off (Lee-Smith 2010). Socio-economic benefits of wastewater irrigation are widely documented (Hanjra et al. 2015a; Makoni et al. 2016). For example, in **Moshi Municipality, Tanzania**, the use of treated wastewater in urban agriculture improved incomes and provides employment. However, improperly practised effluent irrigation is associated with public health risks to workers. Despite this, it still has positive social and economic implications and wastewater irrigation practitioners continue to do it (Kihila et al. 2014). Data from Dar-es-Salaam for the period from 1992 to 2005 (Drechsel and Dongus 2009) showed that total production areas are relatively stable. In recent times, crop production in urban open spaces by residential suburban cultivators in Dar-es-Salaam appears to be a market-driven, highly productive and profitable business activity (Owens 2016). However, the common use of polluted water limits official support for irrigated urban farming. Farmers in **Yaoundé, Cameroon** using wastewater irrigation can sell vegetables in the dry season at double the price of wet season production, and incomes were nearly 50% above the minimum wage, with leafy vegetables providing 8% of protein and 40% of calcium intake of all urban consumers. In Dar-es-Salaam, 67% of farmers had higher than average incomes, with 90% of leafy vegetables and 60% of milk consumed coming from urban and peri-urban agriculture. All crop farmers in Addis Ababa had incomes well above the median national income.

Integrating urban agriculture into urban planning can enhance the benefits of wastewater irrigation in urban and peri-urban areas. Examples include 'green zones' for horticulture in Maputo city, while the city of Ndola in Zambia has recognised crop and livestock production as legitimate land uses in its strategic plan. Many other cities have responded with policy initiatives (Table 2).

Table 2 Policy initiatives supporting urban agriculture in selected African cities

City, country	Policy
Maputo, Mozambique	Long recognised urban agriculture and supported 'green zones' for horticulture
Ndola, Zambia	Recognised crop and livestock production as legitimate land uses in its strategic plan
Lilongwe, Malawi	The 2030 Urban Plan recognises urban agriculture as regulated land use, although raw wastewater use is still against the policy framework in Malawi
Gaborone, Botswana	Established a national incubator providing training to farmers across the country on wastewater irrigation for vegetable production and business skills development
Dar-es-Salaam, Tanzania	Urban agriculture built into Agriculture and Livestock Policy and official planning and management; city provides improved extension services to help build farmers' organisational capacity
Kampala, Uganda	City Council has a Department of Agriculture and provides extension services to farmers, by-laws governing urban agriculture, a typology, and longitudinal measurement of its scale and extent (best practice)
Nairobi, Kenya	Urban agriculture incorporated into national land policy, adopted by parliament in 2010; municipal councils developed by-laws; farmers formed a gender-balanced network
Accra, Ghana	Due to the decentralisation of agricultural ministry, there is also a Directorate for Farming in Accra, providing extension services and support to urban farmers on learning how to organise, advocate, and increase their efforts towards land security, but also tackle issues in accessing safe water for irrigation

Source Authors

4.3 *Big Data*

Structural transformation and nutrition: Analysis of multi-year data for 29 developing countries shows that structural transformation increases total income, and that poverty falls faster with stronger support for agriculture (Webb and Block 2012). In turn, poverty reduction supports improved nutrition, especially in rural areas. The transformation process must be managed through targeted support for smallholder agriculture.

Drivers of food security: Big Data for small farms, including 13,000 farm households from 93 sites in 17 countries across contrasting agro-ecologies, show that the main drivers of food availability in SSA are crop production, off-farm income and market access. Crop production is the major source (63%) of food availability. Off-farm income contribution ranged from 12% for households without enough food available to 27% for households with sufficient food available. Only three key variables (household size, number of livestock and land area) can predict food availability for 72% of households (Frelat et al. 2016), but market access strongly influenced these linkages. This calls for multisector policy harmonisation, incentives and income diversification, instead of a singular focus on area expansion for agricultural development.

Yield gaps are poverty gaps: Household panel data from 21 regions in eight SSA countries show that poverty gaps are increasing with yield gaps, particularly in low potential areas (Dzanku et al. 2015). Indeed, yield gaps are increasing with expansion of cultivated area into marginal lands. Instead of area expansion, investments in intensification and irrigation development could help to close both yield and poverty gaps.

Urban and peri-urban agriculture: IWMI and the University of California, Berkeley modelled the use of polluted water in farming on a global scale. Study results show that 65% of all irrigated areas less than 40 km downstream of urban centres—about 35.9 million ha worldwide—are affected by wastewater flows (Thebo et al. 2017). Of this total area, 29.3 million ha is in countries with very limited wastewater treatment; thus, wastewater reuse provides economic opportunities for smallholders, but exposes 885 million urban consumers, farmers and food vendors to health risks. This calls for urgent investments to enhance the recovery of water, nutrients and energy from wastewater for safe reuse, thus transforming urban wastewater into an economic asset (Hanjra et al. 2015b; Miller et al. 2017). These investments will also improve public health, consumer safety and food handling.

4.4 Telecoupling and Irrigation Development

Water development and irrigation investments increase the interconnectedness of hydrologic and socioeconomic systems and can have cascading effects known as 'telecoupling', i.e. socioeconomic and environmental interactions in coupled human and natural systems at different scales over great distances (Liu et al. 2013). Telecoupling in water development can significantly influence the outcomes and sustainability of development projects. For example, Ethiopia receives about half of its annual budget from foreign development assistance. Water development is linking its hydrology with distant communities and markets, creating new flows of people, materials and investments. This is resulting in cascading impacts and feedback between urbanisation, the economy and the water-food-energy nexus in East Africa (Chignell and Laituri 2016).

Policy changes in leading global economies (USA, China, Brazil and India) will have collateral effects in vulnerable countries in Africa. For example, China in a bid to improve trade to bridge the gap between food production and consumption is supporting **large water transfer projects** in some African countries (for example, Botswana, Namibia, Lesotho and South Africa), and the world's largest and longest one, the South-North Water Transfer Scheme in China with a planned investment of USD 77 billion (Liu et al. 2013). Other examples include the growing demand for **biofuels** and large-scale agricultural land acquisitions which may take water away from human food systems, (Williams et al. 2012; Schoneveld 2014), **conservation investments** such as payments for ecosystem services and rising global **food trade.**

Asian irrigation and the development of West African irrigation is strengthening interconnections between humans and nature (Im et al. 2014; de Vrese et al. 2016). Irrigation development changes agrifood systems over large distances, with spillover effects on food security and land use dynamics. Smallholder farmers in Africa are not just **beneficiaries** of irrigation development through satisfaction of their own food security. They are also **agents** of change, playing a significant role in cumulative irrigation development and influencing complex drivers that transcend spatial, institutional and temporal scales (Table 3).

Table 3 What smallholders contribute to irrigation development and transformational change

Contribution	Remarks
Investments	African farmers invest significantly in irrigation development (capital, labour, canals, pumps, machines, infrastructure, management) beyond official statistics
Innovations	Production practices, site selection, technology adoption/copy, new investment patterns, rural–urban market linkages
Social networks	Farmer interactions with outside agents (informal trading networks, agrodealers, pump mechanics, mobile phone and Internet, contractors, extension, engineers, authorities, civil society, donors) promoting change
Markets	City and regional markets, cash, credit, output and input linkages, transport, banking, finance
Resource management	Sustainable management of land, water, energy and food, despite land tenure risk
Infrastructure	Maintenance of irrigation schemes, canals, roads, bridges, schools, health centres
Knowledge	Indigenous knowledge, ancient water management practices, landraces, customs, traditions, risk management strategies
Youth engagement	Youth engagement in agriculture; youth retention in traditional artisan communities; youth training to provide future generation of irrigators
Human migration	Reduced rural–urban out-migration due to irrigation development; employment of young migrants (e.g. from Burkina Faso in Accra's urban agriculture using wastewater irrigation; farmworkers returning from Zimbabwe to central Mozambique's mountainous irrigation area); labour mobility to large-scale irrigation development schemes and population settlement in uninhabited areas
Food trade	Increased market sale; supply contracts with supermarkets; informal regional market links; pan-African food trade

Source Authors

5 Discussion: Transforming Investments to Feed the Future

> There is no reason why Africa cannot be self-sufficient when it comes to food. It has sufficient arable land. What's lacking is the right seeds, the right irrigation, but also the kinds of institutional mechanisms that ensure that a farmer is going to be able to grow crops, get them to market, get a fair price.
>
> US President Barack Obama, G8, Italy, 10 July 2009. (Cited in Lankford 2009: 476)

5.1 *Investing in Small-Scale Irrigation and Complementary Infrastructure*

A review of 104 studies (82% of them from Africa) indicates that enhancing future food security will require a primary focus on sustainable intensification of African smallholder farming systems along five domains: productivity, economics, environment, social and human well-being (nutrition and social equity). **Strong metrics exist for all domains except social and human well-being which have major gaps** (Smith et al. 2017). Gains in smallholder productivity and poverty reduction are far greater when irrigation investments are combined with complementary interventions in infrastructure—energy, rural roads and rural vehicle supply to ease movement of input and output from farms (Tamene and Megento 2017)—and in services—education and market access—for smallholder irrigators (Hanjra et al. 2009a).

Africa faces unique policy and investment challenges, as smallholders are among the poorest and most food insecure amid droughts and water poverty (Hanjra and Gichuki 2008). For example, about 90% of Africa's arable land is concentrated in just nine countries (Jayne et al. 2014); vast areas are uninhabited and utilisation of available arable land is limited due to lack of investment (Chamberlin et al. 2014); and population is clustered in some areas with unsustainable intensification (Jayne et al. 2014). Livelihood opportunities outside the farming sector must improve to create faster growth in rural non-farm employment (Ricker-Gilbert et al. 2014), along with better population planning and policies (Headey and Jayne 2014).

Africa faces the largest food gap, with its cereal demand tripling by 2050. This will require sustainable intensification, including significant increase in yield, cropping intensity and sustainable expansion of irrigated production (van Ittersum et al. 2016), and water-smart agricultural practices (Nicol et al. 2015). Data spanning 43 years show that climate-smart agriculture in Nigeria will require more area under irrigation to enhance the development of all sub-sectors of agriculture for future food security (Olayide et al. 2016).

5.2 Groundwater Development Policy

A boom in African groundwater utilization is a pre-requisite for irrigation development and improved well-being (Villholth 2013). Africa is nowhere near the level of groundwater use in agriculture in Asia. For example, groundwater accounts for nearly 50% of irrigated area in India (Giordano 2009) whereas it is a relatively underutilised resource in SSA, providing less than 7% of irrigation water. Investments need to be directed particularly towards (i) estimating availability of shallow groundwater to identify high-yielding local sites suited to smallholder irrigation development (Ebrahim and Villholth 2016); and (ii) providing access to affordable energy sources, to drill boreholes and pump water for irrigation (Villholth et al. 2013). Transition to groundwater seems feasible for the **sustainable intensification** of irrigated agriculture, notwithstanding the complex bottlenecks such as incomplete knowledge of groundwater availability, high energy costs, poor market and infrastructure and acute seasonal labour shortages (Amjath-Babu et al. 2016). The strategic importance of groundwater for global water and food security will likely intensify under climate change adaptation strategies (Taylor et al. 2013). Groundwater acts as a buffer against impacts of climate variability and alleviates poverty in low-income settings by reducing crop failure and increasing yields and incomes (Richts and Vrba 2016) and hydroclimatic extremes such as droughts and floods (Pavelic et al. 2015). In Oman, modern groundwater irrigation methods, such as drip irrigation for vegetable production, enhance crop water productivity and economic returns (Al-Said et al. 2012). However, costs and investments needed for groundwater development and pumping are prohibitively high; requiring public financing and cost-sharing business models targeting smallholders (Gebregziabher et al. 2013). For instance, in the Saiss plain in **Morocco**, the boom in groundwater economy benefited entrepreneurial and more affluent farmers, but their greater access to capital and state subsidies contributed to the marginalisation of smallholders, increasing socio-economic inequalities despite rapid transformation (Ameur et al. 2017).

Groundwater policy interventions must target different issues at different scales—local, national and river basin—and involve both formal and informal actors. A systematic review of 37 studies of pumped irrigation systems in 13 countries found: eight countries where motorised pumps are used (**Mauritania, Ethiopia, Nigeria, Mali, Niger, Kenya, South Africa, Malawi**), four where treadle pumps are used (**Malawi, Zimbabwe, Ghana, Kenya**), two where solar pumps are used (**Benin, Ethiopia**) and one where rope pump is used (**Zimbabwe**), but none using wind pumps (Kamwamba-Mtethiwa et al. 2016). The choice of pump and energy is important and must take advantage of recent advances in science and technology. Also, groundwater development policy must be inclusive. Three case studies in **Morocco, Tunisia** and **Algeria** showed that, although groundwater supply chain actors are often informal and operate at the margin of public policy, they are catalysers for groundwater development and their greater involvement can reduce risks, facilitate access to credit and subsidies and dissemination of innovations in groundwater development (Lejars et al. 2017).

5.3 *Investing in Solar Energy for Groundwater Pumping for Irrigation*

Investment in solar energy is needed to promote small-scale distributed irrigation across SSA, where feasible. Mobisol, a German solar company working in Tanzania in partnership with **USAID's Power Africa initiative,** provides solar panels to enable customers use power welding and pipe-cutting equipment, water pumps, egg incubators and fans to make cooking stoves more fuel-efficient. **Scaling Solar**, an initiative of the World Bank, supported an auction in Zambia to introduce and procure solar energy quickly and at very competitive prices—6.02 US cents per kWh, compared to oil-based power which can be three to four times as expensive (Brookings 2016). **Madagascar** and **Senegal** are also participating in a Scaling Solar initiative that could amount to nearly USD 1 billion of investment. **New Deal on Energy in Africa**, a multi-billion-dollar initiative by AfDB, aims to establish 75 million new off-grid connections. Since the launch of USAID's Power Africa in 2013, the programme's off-grid partners have added about 2.5 million new connections (Brookings 2016).

Investment in solar energy in Africa will support agricultural transformation and enhance food security. Solar irrigation has positive effects on labour saving, crop yield and profits on smallholder farms in Ethiopia. Investment in solar technology and upstream pumping for large-scale land reclamation projects in the **Sinai desert, Egypt** and solar energy farms in the Israel desert (Fischhendler et al. 2016) provide useful examples that could revolutionise agriculture in **MENA** and dry regions of SSA. The declining cost of solar panels could expand groundwater use in agriculture. Yet, for uptake by smallholders, the capital cost of solar irrigation pumps (USD 650–3000) must become competitive in comparison to fuel/electric pumps (USD 300–1500) or treadle pumps (USD 25–100). India, for example, has over 11.5 million electric and 6.7 million diesel pumps, but only 2000 solar irrigation pumps (Shah et al. 2014).

New solar pump policy for irrigation should enhance solar power for food security. This requires a supportive business model that packages **conditional subsidies**—linked to compliance with design regulations, crop water requirements and regular monitoring and reporting and **targeted subsidies** for smallholders that allow only smaller solar irrigation pumps (1.5–2.5 kWp) for groundwater irrigation. Solar water pumps could be widely adopted with policy support (subsidy, tax exemption, access to technologies, finance and markets) to address initial affordability problems and facilitate solar power concentration storage in batteries. A solar irrigation policy package, comprising capital cost subsidy and guaranteed buyback of surplus power, could transform both the energy and groundwater economies. The energy payback period would be shorter and surplus solar energy buyback income would encourage farmers to raise the productivity of both groundwater and solar energy, by investing in micro-irrigation and choosing crops with high returns on irrigation. Solar irrigation is aligned with government policy priorities in Ethiopia, Jordan, Morocco, Egypt and Zambia. In off-grid areas, supportive policies are needed to co-optimise solar panels, batteries and water pumps. Solar irrigation saves on fuel costs, by replacing petrol and diesel while also earning revenue from charging mobile phones—at ETB 2 (7 US cents) per phone—and selling water to neighbouring farmers. Decentralised solar power generation in remote villages and rural irrigation schemes could enhance local food security.

5.4 *Investment in Peri-urban and Urban Agriculture*

To support peri-urban and urban agriculture and promote safe reuse of wastewater, policy must address the current mismatch between national food security policies and urban bias, e.g. maize policy in Zambia (Hanjra and Culas 2011). While investment in rural irrigation schemes will remain key to food security, greater attention must be paid to specific challenges in sustainably feeding urban areas. This should involve harmonisation of national food security policy with urban planning that widens the mandate of city councils beyond waste disposal to supporting safe reuse in urban and peri-urban agriculture, in close coordination with irrigation authorities, energy authorities, groundwater management boards, farmers and urban planners (Connor et al. 2017; Hanjra et al. 2017b). To that end, the following key policy implementation strategies are suggested (Hanjra et al. 2017b, d).

- Integrate urban and peri-urban agriculture into national policy processes.
- Provide social incentives and public subsidies on a par with the fertiliser and bioenergy economy, to upscale wastewater irrigation and nutrient reuse in agriculture.
- Mandate estate developers to allocate land parcels for community gardens and residential complexes to undertake onsite wastewater treatment for reuse in order to boost local food production and promote social cohesion.

5.5 *No-Regret Investments for Managing Social Change*

The nature of social change and associated vulnerabilities due to climate change are not well understood (Nelson et al. 2016). Scenario exercises for 2030 conducted by IWMI in East Africa, involving national policy experts and regional stakeholders, show how the relative usefulness of capacity development approaches compared to impact approaches to adaptation planning differs with the level of uncertainty and associated lead time (Vermeulen et al. 2013). **Capacity development approaches** are important for incremental adaptation and innovation, through institutional support to farmers (e.g. financial services education and participatory rehabilitation of existing irrigation schemes) that are feasible, cost-effective and low-risk response.

Impact approaches, involving wholesale reconfiguration of food and farming systems, large-scale anticipatory investments in irrigation and other infrastructure, livelihood diversification and population migration to better endowed areas, are particularly important for transformative adaptation. To avoid disruptive social change, such innovations require strategic guidance on agricultural water management and water policy in order to to make long-term **no-regret interventions**. These could include investments in wastewater reuse for irrigation; wider storage options beyond surface reservoirs, including groundwater as a buffer against long-term droughts (Scanlon and Vladimir 2016); micro-irrigation technologies; rainwater harvesting; farmer gene banks (e.g. ICARDA's global seed vault in a remote island of Norway); indigenous crop races; and cultivation of wild foods and hardy land races under rainfed systems. This would help to prepare against catastrophic risks and safeguard food security and sustainability into the distant future.

6 Conclusions and Outlook

This chapter presented evidence on irrigation and food security linkages across Africa and produced three major conclusions. Firstly, investments in irrigation contribute to poverty reduction. Here, 'irrigation' refers to surface water and groundwater use in rural irrigation schemes, as well as wastewater use for food production in rural, peri-urban and urban areas. Secondly, strong evidence exists that investments in irrigation enhance food security. Thirdly, existing evidence is supportive but still insufficient to draw broader conclusions on nutrition outcomes, primarily because nutrition is only now being considered as an explicit objective of irrigation development and agriculture policy (Pingali 2015). Nutrition-sensitive irrigation programmes and delivery platforms are therefore needed to help realise the full potential of irrigation for enhancing both food and nutrition security (Hanjra et al. 2017c). Strategic priorities to enhance food and nutrition security are investments in smallholder rural irrigation schemes, peri-urban and urban agriculture, and related support measures, including rural infrastructure and solar energy for well-distributed irrigation development.

There is a need to frame 'irrigation' to integrate socio-economic and environmental interactions affecting sustainability, across local to global levels, through telecoupling. **System integration and sustainability** can transform how policymakers think about irrigation and agricultural water management and facilitate the training of a new generation that is well-equipped to develop food security and environmental sustainability solutions. There is also a need to realise the productive function of urban and peri-urban agriculture within urban planning, and a need for better integration of food and nutrition security issues in land use planning, especially within wastewater sanitation systems in cities and towns across Africa.

Acknowledgements This work was supported by the CGIAR Research Programme on Water, Land and Ecosystems (WLE), led by IWMI. The work was presented at the European Commission

Joint Research Centre (JRC) workshop, held in November 2015 in Seville, Spain. We gratefully acknowledge the constructive feedback and guidance of workshop participants which helped to improve the work, and the financial support from JRC Seville, Spain, to the first author to attend the workshop.

Appendix: Indicators of Regional Change and Transformation in SSA

World Bank data showed that, over the period 1990–2013, population growth remained high (2.7% per annum). Agricultural productivity growth is continuing but not apace to feed the population. Majority of the poor (82%) still live in rural areas and the majority of rural households (69%) earn income from agriculture. The share of agriculture in employment (61%) and GDP (25%) is still high resulting in widespread poverty (417 million people) across major agro-ecological zones in Africa. Real agricultural value added has been growing (4.1%) over the period 1990–2013, but this was countered by high population growth and grew only slowly (1.4%) in per capita terms (Barrett et al. 2017). Food security has improved, largely due to annual growth in cultivated area under cereals (1.3%), but cereal yield growth (1.6%) has been far lower than in Asia during the Green Revolution. Africa yield levels started from a very low base and remain low. For example, average cereal yield in Africa today (about 1.5 t/ha) is less than half of the level in South Asia (3.1 t/ha), about a quarter of the level in China (6 t/ha) (Barrett et al. 2017), and about one eighth of the level in Australia.

Structural transformation in Africa has been towards low productivity, non-tradable services in urban areas, rather than tradable manufacturing (Rodrik 2016). This has ignited urbanisation and the emergence of consumption cities (Gollin et al. 2016), with large metabolic throughput of water, food and energy and a rising share of food imports to ensure local food security. Such resource-driven urbanisation and structural transformation has led to expansion in slums, poor water and sanitation services, a widening rural-urban income gap and inequality (Gollin et al. 2016).

References

Abro, Z. A., Alemu, B. A., & Hanjra, M. A. (2014). Policies for agricultural productivity growth and poverty reduction in rural Ethiopia. *World Development, 59,* 461–474.

AfDB. (2016). *Feed Africa—Strategy for agricultural transformation in Africa: 2016–2025.* African Development Bank.

Alaofè, H., Burney, J., Naylor, R., & Taren, D. (2016). Solar-powered drip irrigation impacts on crops production diversity and dietary diversity in Northern Benin. *Food and Nutrition Bulletin, 37,* 164–175.

Al-Said, F. A., Ashfaq, M., Al-Barhi, M., Hanjra, M. A., & Khan, I. A. (2012). Water productivity of vegetables under modern irrigation methods in Oman. *Irrigation and Drainage, 61,* 477–489.

Ameur, F., Kuper, M., Lejars, C., & Dugué, P. (2017). Prosper, survive or exit: Contrasted fortunes of farmers in the groundwater economy in the Saiss plain (Morocco). *Agricultural Water Management, 191,* 207–217.

Amjath-Babu, T. S., Krupnik, T. J., Kaechele, H., Aravindakshan, S., & Sietz, D. (2016). Transitioning to groundwater irrigated intensified agriculture in Sub-Saharan Africa: An indicator based assessment. *Agricultural Water Management, 168,* 125–135.

Barrett, C. B., Christiaensen, L., Sheahan, M. B., & Shimeles, A. (2017). *On the structural transformation of rural Africa* (Policy Research Working Paper 7938). Washington, DC: The World Bank.

Bren d'Amour, C., Reitsma, F., Baiocchi, G., Barthel, S., Güneralp, B., Erb, K.-H., et al. (2016). Future urban land expansion and implications for global croplands. *Proceedings of the National Academy of Sciences.*

Brookings. (2016). *The US-Africa business forum: Investing in solar energy.* Retrieved January 15, 2016, from https://www.brookings.edu/blog/africa-in-focus/2016/09/15/the-us-africa-business-forum-investing-in-solar-energy/.

Burney, J. A., & Naylor, R. L. (2012). Smallholder irrigation as a poverty alleviation tool in Sub-Saharan Africa. *World Development, 40,* 110–123.

Burney, J., Woltering, L., Burke, M., Naylor, R., & Pasternak, D. (2010). Solar-powered drip irrigation enhances food security in the Sudano-Sahel. *Proceedings of the National Academy of Sciences of the United States of America, 107,* 1848–1853.

Carletto, G., Ruel, M., Winters, P., & Zezza, A. (2015). Farm-level pathways to improved nutritional status: Introduction to the special issue. *The Journal of Development Studies (Special Issue), 51,* 945–957.

Chamberlin, J., Jayne, T. S., & Headey, D. (2014). Scarcity amidst abundance? Reassessing the potential for cropland expansion in Africa. *Food Policy, 48,* 51–65.

Chignell, S. M., & Laituri, M. J. (2016). Telecoupling, urbanization, and the unintended consequences of water development aid in Ethiopia. *Geological Society of America Special Papers, 520,* 125–135.

Comas, J., Connor, D., Isselmou, M. E. M., Mateos, L., & Gómez-Macpherson, H. (2012). Why has small-scale irrigation not responded to expectations with traditional subsistence farmers along the Senegal River in Mauritania? *Agricultural Systems, 110,* 152–161.

Conceição, P., Levine, S., Lipton, M., & Warren-Rodríguez, A. (2016). Toward a food secure future: Ensuring food security for sustainable human development in Sub-Saharan Africa. *Food Policy, 60,* 1–9.

Connor, R., Renata, A., Ortigara, C., Koncagül, E., Uhlenbrook, S., Lamizana-Diallo, B., et al. (2017) *The United Nations world water development report 2017. Wastewater: The untapped resource.* New York, NY.

de Vrese, P., Hagemann, S., & Claussen, M. (2016). Asian irrigation, African rain: Remote impacts of irrigation. *Geophysical Research Letters.* https://doi.org/10.1002/2016gl068146.

Demont, M., Fiamohe, R., & Kinkpé, A. T. (2017). Comparative advantage in demand and the development of rice value chains in West Africa. *World Development.* https://doi.org/10.1016/j.worlddev.2017.04.004.

Dillon, A. (2011). The effect of irrigation on poverty reduction, asset accumulation, and informal insurance: Evidence from Northern Mali. *World Development, 39,* 2165–2175.

Dixon, J., Garrity, D., Boffa, J.-M., Williams, T., Amede, T., Auricht, C., et al. (Eds.). (2017). *Farming systems and food security in Africa: Priorities for science and policy under global change.* London and New York, NY: Routledge.

Domènech, L. (2015). Improving irrigation access to combat food insecurity and undernutrition: A review. *Global Food Security, 6,* 24–33.

Drechsel, P., & Dongus, S. (2009). Dynamics and sustainability of urban agriculture: Examples from Sub-Saharan Africa. *Sustainability Science, 5,* 69.

Drechsel, P., & Keraita, B. (Eds.). (2014). *Irrigated urban vegetable production in Ghana: Characteristics, benefits and risk mitigation* (2nd ed., 247 pp). Colombo, Sri Lanka: International Water Management Institute (IWMI).

Dzanku, F. M., Jirström, M., & Marstorp, H. (2015). Yield gap-based poverty gaps in rural Sub-Saharan Africa. *World Development, 67,* 336–362.

Ebrahim, G. Y., & Villholth, K. G. (2016). Estimating shallow groundwater availability in small catchments using streamflow recession and instream flow requirements of rivers in South Africa. *Journal of Hydrology, 541,* 754–765.

Fan, S., Hazell, P., & Haque, T. (2000). Targeting public investments by agro-ecological zone to achieve growth and poverty alleviation goals in rural India. *Food Policy, 25,* 411–428.

Fan, S., Gulati, A., & Thorat, S. (2008). Investment, subsidies, and pro-poor growth in rural India. *Agricultural Economics, 39,* 163–170.

Fanadzo, M. (2012). Revitalisation of smallholder irrigation schemes for poverty alleviation and household food security in South Africa: A review. *African Journal of Agricultural Research, 7,* 1956–1969.

FAO. (2012). *Growing greener cities in Africa. First status report on urban and peri-urban horticulture in Africa.* Rome: Food and Agriculture Organization of the United Nations.

FAO. (2015). *Regional overview of food insecurity: African food security prospects brighter than ever.* Accra, Ghana: Food and Agriculture Organization of the United Nations.

FAO. (2019). *The state of food security and nutrition in the world. Safeguarding against economic slowdowns and downturns.* Rome: Food and Agriculture Organization of the United Nations, Rome.

Fiorella, K. J., Chen, R. L., Milner, E. M., & Fernald, L. C. H. (2016). Agricultural interventions for improved nutrition: A review of livelihood and environmental dimensions. *Global Food Security, 8,* 39–47.

Fischhendler, I., Boymel, D., & Boykoff, M. T. (2016). How competing securitized discourses over land appropriation are constructed: The promotion of solar energy in the Israeli Desert. *Environmental Communication, 10,* 147–168.

Frelat, R., Lopez-Ridaura, S., Giller, K. E., Herrero, M., Douxchamps, S., Djurfeldt, A. A., et al. (2016). Drivers of household food availability in Sub-Saharan Africa based on big data from small farms. *Proceedings of the National Academy of Sciences, 113,* 458–463.

Gebregziabher, G., Villholth, K. G., Hanjra, M. A., Yirga, M., & Namara, R. E. (2013). Cost-benefit analysis and ideas for cost sharing of groundwater irrigation: Evidence from north-eastern Ethiopia. *Water International, 38,* 852–863.

Gerland, P., Raftery, A. E., Ševčíková, H., Li, N., Gu, D., Spoorenberg, T., et al. (2014). World population stabilization unlikely this century. *Science, 346,* 234–237.

Giordano, M. (2009). Global groundwater? Issues and solutions. *Annual Review of Environment and Resources, 34,* 153–178.

Giordano, M., Turral, H., Scheierling, S. M., Tréguer, D. O., & McCornick, P. G. (2017). *Beyond 'more crop per drop'* (Research Report 169). Colombo, Sri Lanka: International Water Management Institute (IWMI); Washington, DC: The World Bank.

Gollin, D., Jedwab, R., & Vollrath, D. (2016). Urbanization with and without industrialization. *Journal of Economic Growth, 21,* 35–70.

GoM. (2013). Urban structure plan of Lilongwe City. Lilongwe, Malawi: Government of Malawi (GoM).

Grewal, S. S., & Grewal, P. S. (2012). Can cities become self-reliant in food? *Cities, 29,* 1–11.

Grey, D., & Sadoff, C. W. (2007). Sink or swim? Water security for growth and development. *Water Policy, 9,* 545–571.

Hagos, F., & Mamo, K. (2014). Financial viability of groundwater irrigation and its impact on livelihoods of smallholder farmers: The case of eastern Ethiopia. *Water Resources and Economics, 7,* 55–65.

Hagos, F., Mulugeta, A., Erkossa, T., Langan, S., Lefore, N., & Abebe, Y. (2017). Poverty profiles and nutritional outcomes of using spate irrigation in Ethiopia. *Irrigation and Drainage.*

Hanjra, M. A., & Culas, R. J. (2011). The political economy of maize production and poverty reduction in Zambia: Analysis of the last 50 years. *Journal of Asian and African Studies, 46,* 546–566.

Hanjra, M. A., & Gichuki, F. (2008). Investments in agricultural water management for poverty reduction in Africa: Case studies of Limpopo, Nile, and Volta river basins. *Natural Resources Forum, 32,* 185–202.

Hanjra, M. A., Ferede, T., & Gutta, D. G. (2009a). Pathways to breaking the poverty trap in Ethiopia: Investments in agricultural water, education, and markets. *Agricultural Water Management, 96,* 1596–1604.

Hanjra, M. A., Ferede, T., & Gutta, D. G. (2009b). Reducing poverty in Sub-Saharan Africa through investments in water and other priorities. *Agricultural Water Management, 96,* 1062–1070.

Hanjra, M. A., Blackwell, J., Carr, G., Zhang, F., & Jackson, T. M. (2012). Wastewater irrigation and environmental health: Implications for water governance and public policy. *International Journal of Hygiene and Environmental Health, 215,* 255–269.

Hanjra, M. A., Drechsel, P., Mateo-Sagasta, J., Otoo, M., & Hernandez-Sancho, F. (2015a). Assessing the finance and economics of resource recovery and reuse solutions across scales (Chap. 7). In P. Drechsel, M. Qadir, & D. Wichelns (Eds.), *Wastewater: Economic asset in an urbanizing world.* Dordrecht, Heidelberg, New York, London: Springer.

Hanjra, M. A., Drechsel, P., Wichelns, D., & Qadir, M. (2015b). Transforming urban wastewater into an economic asset: Opportunities and challenges (Chap. 14). In P. Drechsel, M. Qadir, & D. Wichelns (Eds.), *Wastewater: Economic asset in an urbanizing world.* Dordrecht, Heidelberg, New York, London: Springer.

Hanjra, M. A., Kayumbe, J. R. L., Barron, J., & Williams, T. (2016a, January). *Drivers of successful small scale irrigation schemes in Tanzania* (Draft Report). IWMI/WLE project on Small Scale Irrigation Schemes in Sub-Saharan Africa. Colombo, Sri Lanka: International Water Management Institute.

Hanjra, M. A., Zawe, C., Barron, J., & Williams, T. (2016b, January). *Public private partnership investments in irrigation and sustainable agricultural water management solutions, Zimbabwe* (Draft Report). IWMI/WLE project on Small Scale Irrigation Schemes in Sub-Saharan Africa. Colombo, Sri Lanka: International Water Management Institute.

Hanjra, M. A., Drechsel, P., & Masundire, H. M. (2017a). Urbanization, water quality and water reuse in the Zambezi river basin (Chap. 7). In J. Lautze, Z. Phiri, & V. Smakhtin (Eds.), *Zambezi river basin: Water and sustainable development.* New York: Earthscan; Colombo: IWMI (Forthcoming).

Hanjra, M. A., Lydecker, M., Drechsel, P., & Paul, J. (2017b). Rural-urban food and nutrient dynamics and nutrient recovery from waste in developing countries. In J. Zeunert & T. Waterman (Eds.), *The Routledge handbook of landscape and food.* Routledge: London, UK, New York, NY.

Hanjra, M. A., Noble, A., Langan, S., & Lautze, J. (2017c). Feeding the 10 billion within the sustainable development goals framework (Chap. 2). In I. J. Gordon, H. H. T. Prins, & G. R. Squire (Eds.), *Food production and nature conservation: Conflicts and solutions.* London, UK, New York, NY: Earthscan (Routledge).

Hanjra, M. A., Qadir, M., & Drechsel, P. (2017d). Cross-sectoral resource recovery and reuse in Asia and Africa: Opportunities for a circular metabolism. In W. Montasser (Ed.), *Springer book series: RESILIENT CITIES: Re-thinking Urban Transformation: Circular urban metabolism: Generating co-benefits through urban resilience transition.* London, UK: Springer (Forthcoming).

Hanjra, M. A., Wichelns, D., & Drechsel, P. (2017e). Investing in water management in rural and urban landscapes to achieve and sustain global food security. In J. Zeunert & T. Waterman (Eds.), *The Routledge handbook of landscape and food.* Routledge: London, UK, New York, NY.

Headey, D. D., & Jayne, T. S. (2014). Adaptation to land constraints: Is Africa different? *Food Policy, 48,* 18–33.

Holm, R., Wanda, E., Kasulo, V., & Gwayi, S. (2014). Identification of the potential opportunities, barriers, and threats within the sector in taking up sanitation as a business: Rural sanitation in Nkhata Bay District (Malawi). *Waterlines, 33,* 269–274.

Hosu, S. Y., Cishe, E., & Luswazi, P. (2016). Vulnerability to climate change in the Eastern Cape province of South Africa: What does the future holds for smallholder crop farmers? *Agrekon, 55,* 133–167.

Hussain, I., & Hanjra, M. A. (2003). Does irrigation water matter for rural poverty alleviation? Evidence from South and South-East Asia. *Water Policy, 5,* 429–442.

Hussain, I., & Hanjra, M. A. (2004). Irrigation and poverty alleviation: Review of the empirical evidence. *Irrigation and Drainage, 53,* 1–15.

IFAD. (2011). *Rural poverty report 2011* [Online]. International Fund for Agricultural Development (IFAD).

Im, E.-S., Marcella, M. P., & Eltahir, E. A. B. (2014). Impact of potential large-scale irrigation on the West African monsoon and its dependence on location of irrigated area. *Journal of Climate, 27,* 994–1009.

Jayne, T. S., Chamberlin, J., & Headey, D. D. (2014). Land pressures, the evolution of farming systems, and development strategies in Africa: A synthesis. *Food Policy, 48,* 1–17.

Kamwamba-Mtethiwa, J., Weatherhead, K., & Knox, J. (2016). Assessing performance of small-scale pumped irrigation systems in Sub-Saharan Africa: Evidence from a systematic review. *Irrigation and Drainage, 65,* 308–318.

Katic, P. G., Namara, R. E., Hope, L., Owusu, E., & Fujii, H. (2013). Rice and irrigation in West Africa: Achieving food security with agricultural water management strategies. *Water Resources and Economics, 1,* 75–92.

Keats, S., & Wiggins, S. (2017). *Future diets: Implications for agriculture and food prices* (Cancer/ODI Report). London, UK.

Kihila, J., Mtei, K., & Njau, K. (2014). Wastewater treatment for reuse in urban agriculture; the case of Moshi Municipality, Tanzania. *Physics and Chemistry of the Earth, 72,* 104–110.

Kurosaki, T. (2003). Specilization and diversification in agricultural transformation: The case of West Punjab, 1903–1992. *American Journal of Agricultural Economics, 85,* 372–386.

Lejars, C., Daoudi, A. & Amichi, H. (2017). The key role of supply chain actors in groundwater irrigation development in North Africa. Hydrogeology Journal 25, 1593–1606.

Lankford, B. (2009). Viewpoint—The right irrigation? Policy directions for agricultural water management in Sub-Saharan Africa. *Water Alternatives, 2,* 476–480.

Larson, D. F., Muraoka, R., & Otsuka, K. (2016). Why African rural development strategies must depend on small farms. *Global Food Security, 10,* 39–51.

Leah, H. S., James, S. G., Navin, R., Mario, H., & Paul, C. W. (2016). Subnational distribution of average farm size and smallholder contributions to global food production. *Environmental Research Letters, 11,* 124010.

Lee-Smith, D. (2010). Cities feeding people: An update on urban agriculture in equatorial Africa. *Environment and Urbanization, 22,* 483–499.

Liu, J., Hull, V., Batistella, M., DeFries, R., Dietz, T., Fu, F., et al. (2013). Framing sustainability in a telecoupled world. *Ecology and Society, 18.*

Lowder, S. K., Skoet, J., & Raney, T. (2016). The number, size, and distribution of farms, smallholder farms, and family farms worldwide. *World Development.*

Lydecker, M., & Drechsel, P. (2010). Urban agriculture and sanitation services in Accra, Ghana: The overlooked contribution. *International Journal of Agricultural Sustainability, 8,* 94–103.

Magombeyi, M. S., Taigbenu, A. E., & Barron, J. (2016). Rural food insecurity and poverty mappings and their linkage with water resources in the Limpopo River Basin. *Physics and Chemistry of the Earth, Parts A/B/C, 92,* 20–33.

Makoni, F. S., Thekisoe, O. M. M., & Mbati, P. A. (2016). Urban wastewater for sustainable urban agriculture and water management in developing countries. In T. Younos & T. E. Parece (Eds.), *Sustainable water management in urban environments.* Cham: Springer International Publishing.

Maziya, M., Mudhara, M., & Chitja, J. (2017). What factors determine household food security among smallholder farmers? Insights from Msinga, KwaZulu-Natal, South Africa. *Agrekon, 56,* 40–52.

Miller, R. L., Anu, R., & Priyanie, A. (2017). Wastewater treatment and reuse in urban agriculture: Exploring the food, energy, water, and health nexus in Hyderabad, India. *Environmental Research Letters, 12,* 075005.

Mishra, A. K., Khanal, A. R., & Mohanty, S. (2017). Gender differentials in farming efficiency and profits: The case of rice production in the Philippines. *Land Use Policy, 63,* 461–469.

Moyo, T., & Machethe, C. L. (2016). The relationship between smallholder irrigation and household food availability and dietary diversity in Greater Tzaneen Municipality of Limpopo Province, South Africa. *Journal of Sustainable Development, 9,* 165–178.

Msilimba, G., & Wanda, E. M. M. (2012). *Wastewater production, treatment, and use in Malawi.* Luwinga, Mzuzu, Malawi: Water and Sanitation Centre of Excellence.

Muchara, B., Ortmann, G., Mudhara, M., & Wale, E. (2016). Irrigation water value for potato farmers in the Mooi River Irrigation Scheme of KwaZulu-Natal, South Africa: A residual value approach. *Agricultural Water Management, 164*(Part 2), 243–252.

Mutengu, S., Hoko, Z., & Makoni, F. S. (2007). An assessment of the public health hazard potential of wastewater reuse for crop production. A case of Bulawayo city, Zimbabwe. *Physics and Chemistry of the Earth, Parts A/B/C, 32,* 1195–1203.

Namara, R. E., Awuni, J. A., Barry, B., Giordano, M., Hope, L., Owusu, E. S., et al. (2011). *Smallholder shallow groundwater irrigation development in the upper east region of Ghana* (Research Report 134). Colombo, Sri Lanka: International Water Management Institute.

Nelson, M. C., Ingram, S. E., Dugmore, A. J., Streeter, R., Peeples, M. A., McGovern, T. H., et al. (2016). Climate challenges, vulnerabilities, and food security. *Proceedings of the National Academy of Sciences, 113,* 298–303.

Nicol, A., Langan, S., Victor, M., & Gonsalves, J. (2015). *Water-smart agriculture in East Africa* (352 pp). Colombo, Sri Lanka: International Water Management Institute (IWMI); CGIAR Research Program on Water, Land and Ecosystems (WLE); Kampala, Uganda: Global Water Initiative East Africa (GWI EA).

Nonvide, G. M. A., Sarpong, D. B., Kwadzo, G. T. M., Anim-Somuah, H., & Amoussouga Gero, F. (2017). Farmers' perceptions of irrigation and constraints on rice production in Benin: A stakeholder-consultation approach. *International Journal of Water Resources Development,* 1–21.

Olayide, O. E., Tetteh, I. K., & Popoola, L. (2016). Differential impacts of rainfall and irrigation on agricultural production in Nigeria: Any lessons for climate-smart agriculture? *Agricultural Water Management, 178,* 30–36.

Olney, D. K., Pedehombga, A., Ruel, M. T., & Dillon, A. (2015). A 2-year integrated agriculture and nutrition and health behavior change communication program targeted to women in Burkina Faso reduces anemia, wasting, and diarrhea in children 3–12.9 months of age at baseline: A cluster-randomized controlled trial. *The Journal of Nutrition, 145,* 1317–1324.

Oni, S. A., Maliwichi, L. L., & Obadire, O. S. (2011). Assessing the contribution of smallholder irrigation to household food security, in comparison to dryland farming in Vhembe district of Limpopo province, South Africa. *African Journal of Agricultural Research, 6,* 2188–2197.

Otoo, M., & Drechsel, P. (Eds.). (2017). *Resource recovery from waste: Business models for energy, nutrients and water reuse in low- and middle-income countries*. Earthscan (in press).

Owens, G. R. (2016). 'We are not farmers': Dilemmas and prospects of residential suburban cultivators in contemporary Dar es Salaam, Tanzania. *Journal of Modern African Studies, 54,* 443–467.

Pandey, V. L., Mahendra Dev, S., & Jayachandran, U. (2016). Impact of agricultural interventions on the nutritional status in South Asia: A review. *Food Policy, 62,* 28–40.

Pavelic, P., Brindha, K., Amarnath, G., Eriyagama, N., Muthuwatta, L., Smakhtin, V., et al. (2015). *Controlling floods and droughts through underground storage: From concept to pilot implementation in the Ganges River Basin* (IWMI Research Report 165) (33 pp). Colombo, Sri Lanka: International Water Management Institute (IWMI). https://doi.org/10.5337/2016.200.

Pingali, P. (2007). Westernization of Asian diets and the transformation of food systems: Implications for research and policy. *Food Policy, 32,* 281–298.

Pingali, P. L. (2012). Green Revolution: Impacts, limits, and the path ahead. *Proceedings of the National Academy of Sciences, 109,* 12302–12308.

Pingali, P. (2015). Agricultural policy and nutrition outcomes—Getting beyond the preoccupation with staple grains. *Food Security, 7,* 583–591.

Rapsomanikis, G. (2015). Small farms big picture: Smallholder agriculture and structural transformation. *Development, 58,* 242–255.

Richts, A., & Vrba, J. (2016). Groundwater resources and hydroclimatic extremes: Mapping global groundwater vulnerability to floods and droughts. *Environmental Earth Sciences, 75,* 1–15.

Ricker-Gilbert, J., Jumbe, C., & Chamberlin, J. (2014). How does population density influence agricultural intensification and productivity? Evidence from Malawi. *Food Policy, 48,* 114–128.

Riesgo, L., Louhichi, K., Gomez y Paloma, S., Hazell, P., Ricker-Gilbert, J., Wiggins, S., et al. (2016). *Food and nutrition security and role of smallholder farms: Challenges and opportunities—Workshop proceedings*. Seville, Spain: European Commission Joint Research Centre. https://doi.org/10.2791/653314.

Rockström, J., Williams, J., Daily, G., Noble, A., Matthews, N., Gordon, L., et al. (2016). Sustainable intensification of agriculture for human prosperity and global sustainability. *Ambio,* 1–14.

Rodrik, D. (2016). Premature deindustrialization. *Journal of Economic Growth, 21,* 1–33.

Sadoff, C. W., Hall, J. W., Grey, D., Aerts, J. C. J. H., Ait-Kadi, M., Brown, C., et al. (2015). *Securing water, sustaining growth: Report of the GWP/OECD task force on water security and sustainable growth* (180 pp). Oxford, UK: University of Oxford.

Sahn, D. E., & Younger, S. D. (2017). The incidence of child health improvements. *Review of Development Economics, 21,* 304–320.

Salami, A., Kamara, A. B., & Brixiova, Z. (2010). *Smallholder agriculture in East Africa: Trends, constraints and opportunities* (African Development Bank Working Paper). Tunis, Tunisia.

Samson, S., Mdegela, R. H., Permin, A., Mlangwa, J. E., & Mahonge, C. P. (2017). Obstacles to low quality water irrigation of food crops in Morogoro, Tanzania. *Journal of Sustainable Development, 10,* 1.

Scanlon, B. R., & Vladimir, S. (2016). Focus on water storage for managing climate extremes and change. *Environmental Research Letters, 11,* 120208.

Schmidt, S., Magigi, W., & Godfrey, B. (2015) The organization of urban agriculture: Farmer associations and urbanization in Tanzania. *Cities, 42*(Part B), 153–159.

Schoneveld, G. C. (2014). The geographic and sectoral patterns of large-scale farmland investments in Sub-Saharan Africa. *Food Policy, 48,* 34–50.

Shah, T., Verma, S., & Durga, N. (2014). Karnataka's smart, new solar pump policy for irrigation. *Economic & Political Weekly, 49,* 11.

Sheahan, M., & Barrett, C. B. (2017). Ten striking facts about agricultural input use in Sub-Saharan Africa. *Food Policy, 67,* 12–25.

Shinkai, N., Sawada, Y., Hussain, I., Hanjra, M. A., Thrikawala, S., & Wijeratne, D. (2007). *Impact of irrigation infrastructure development on dynamics of incomes and poverty: Econometric evidence using panel data from Sri Lanka* (JBIC Research Paper No. 32) [Online]. Tokyo, Japan: Japan Bank for International Cooperation. Online at: http://www.jbic.go.jp/english/research/report/paper/index.php.

Sibhatu, K. T., Krishna, V. V., & Qaim, M. (2015). Production diversity and dietary diversity in smallholder farm households. *Proceedings of the National Academy of Sciences, 112,* 10657–10662.

Sidibé, Y., & Williams, T. O. (2016). Agricultural land investments and water management in the Office du Niger, Mali: Options for improved water pricing. *Water International, 41,* 738–755.

Sinyolo, S., Mudhara, M., & Wale, E. (2014). The impact of smallholder irrigation on household welfare: The case of Tugela Ferry irrigation scheme in KwaZulu-Natal, South Africa. *Water SA, 40.*

Smith, A., Snapp, S., Chikowo, R., Thorne, P., Bekunda, M., & Glover, J. (2017). Measuring sustainable intensification in smallholder agroecosystems: A review. *Global Food Security, 12,* 127–138.

Tamene, S., & Megento, T. L. (2017). The effect of rural road transport infrastructure on smallholder farmers' agricultural productivity in Horro Guduru Wollega Zone, Western Ethiopia. *AUC Geographica, 52,* 79–89.

Taylor, R. G., Scanlon, B., Doll, P., Rodell, M., van Beek, R., Wada, Y., et al. (2013). Ground water and climate change. *Nature Climate Change, 3,* 322–329.

Tedesco, C., Petit, C., Billen, G., Garnier, J., & Personne, E. (2017). Potential for recoupling production and consumption in peri-urban territories: The case-study of the Saclay plateau near Paris, France. *Food Policy, 69,* 35–45.

Thebo, A. L., Drechsel, P., Lambin, E. F., & Nelson, K. L. (2017). A global, spatially-explicit assessment of irrigated croplands influenced by urban wastewater flows. *Environmental Research Letters, 12,* 074008.

Timmer, C. P. (2017). Food security, structural transformation, markets and government policy. *Asia & the Pacific Policy Studies, 4,* 4–19.

Unver, O., Wahaj, R., Lorenzon, E., Mohammedi, K., Osias, J. R., Reinders, F., et al. (2016). *Key and smart actions to alleviate hunger and poverty through irrigation and drainage* (FAO/ICID Background Paper). Online at: http://www.icid.org/wif2_bg_pap_st3.pdf.

van Averbeke, W., Denison, J., & Mnkeni, P. (2011). Smallholder irrigation schemes in South Africa: A review of knowledge generated by the Water Research Commission. *Water SA, 37,* 797–808.

van Ittersum, M. K., van Bussel, L. G. J., Wolf, J., Grassini, P., van Wart, J., Guilpart, N., et al. (2016). Can Sub-Saharan Africa feed itself? *Proceedings of the National Academy of Sciences, 113,* 14964–14969.

Verdier-Chouchane, A., & Boly, A. (2017). Introduction: Challenges to Africa's agricultural transformation—Special issue paper. *African Development Review, 29,* 75–77.

Vermeulen, S. J., Challinor, A. J., Thornton, P. K., Campbell, B. M., Eriyagama, N., Vervoort, J. M., et al. (2013). Addressing uncertainty in adaptation planning for agriculture. *Proceedings of the National Academy of Sciences, 110,* 8357–8362.

Villholth, K. G. (2013). Groundwater irrigation for smallholders in Sub-Saharan Africa—A synthesis of current knowledge to guide sustainable outcomes. *Water International, 38,* 369–391.

Villholth, K. G., Tøttrup, C., Stendel, M., & Maherry, A. (2013). Integrated mapping of groundwater drought risk in the Southern African Development Community (SADC) region. *Hydrogeology Journal, 21,* 863–885.

Ward, F. A., Amer, S. A., & Ziaee, F. (2013). Water allocation rules in Afghanistan for improved food security. *Food Security, 5,* 35–53.

Webb, P., & Block, S. (2012). Support for agriculture during economic transformation: Impacts on poverty and undernutrition. *Proceedings of the National Academy of Sciences, 109,* 12309–12314.

Weldesilassie, A. B., Boelee, E., Drechsel, P., & Dabbert, S. (2011). Wastewater use in crop production in peri-urban areas of Addis Ababa: Impacts on health in farm households. *Environment and Development Economics, 16,* 25–49.

Wichelns, D. (2014). Investing in small, private irrigation to increase production and enhance livelihoods. *Agricultural Water Management, 131,* 163–166.

Wiggins, S., Kirsten, J., & Llambí, L. (2010). The future of small farms. *World Development, 38,* 1341–1348.

Williams, T. O. (2015). Reconciling food and water security objectives of MENA and Sub-Saharan Africa: Is there a role for large-scale agricultural investments? *Food Security, 7,* 1199–1209.

Williams, T. O., Gyampoh, B., Kizito, F., & Namara, R. (2012). Water implications of large-scale land acquisitions in Ghana. *Water Alternatives, 5,* 243–265.

Woodhouse, P., Veldwisch, G. J., Venot, J.-P., Brockington, D., Komakech, H., & Manjichi, Â. (2017). African farmer-led irrigation development: Re-framing agricultural policy and investment? *The Journal of Peasant Studies, 44,* 213–233.

World Bank. (2008). *World development report 2008: Agriculture for development.* Washington, DC: The World Bank.

World Bank. (2018). *Poverty and shared prosperity 2018: Piecing together the poverty puzzle.* Washington, DC: The World Bank.

Zeweld, W., Huylenbroeck, G. V., Hidgot, A., Chandrakanth, M. G., & Speelman, S. (2015). Adoption of small-scale irrigation and its livelihood impacts in northern Ethiopia. *Irrigation and Drainage, 64,* 655–668.

Zezza, A., & Tasciotti, L. (2010). Urban agriculture, poverty, and food security: Empirical evidence from a sample of developing countries. *Food Policy, 35,* 265–273.

5

The Role of Smallholder Farms in a Changing World

Shenggen Fan and Christopher Rue

Abstract Despite progress, multiple burdens of malnutrition persist worldwide: 795 million people are hungry more than 2 billion people suffer from micronutrient deficiencies and over 2 billion are overweight or obese. At the same time, various challenges continue to threaten global food security and nutrition. Smallholder farmers are a key to ending hunger and undernutrition worldwide, but they are increasingly facing barriers to profitability. Yet smallholders should not all receive the same kind of support; they are not a homogenous group. While some smallholders should be supported to move up to commercially oriented and profitable farming systems, some should be supported to move out to seek non-farm employment opportunities. Strategies to promote smallholder agriculture as a business can help to overcome these obstacles and move smallholders with profit potential towards greater prosperity, while also contributing to the achievement of multiple Sustainable Development Goals (SDGs).

1 Introduction

In the coming decades, world agriculture will need to undergo major changes to meet the future food demands of a growing and increasingly rich and urbanised population. Smallholders in developing countries play a key role worldwide in this food security equation. More than 80% (475 million) of the world's farms operate on less than two hectares of land. Although these farms account for only 12% of the world's farmland, they provide an estimated 80% of the food produced in Asia and in sub-Saharan Africa (SSA) (Lowder et al. 2014). Despite the key role smallholder farms play in achieving global food security and nutrition, they are a vulnerable group often neglected by development policy and they account for most of the world's poor and hungry.

S. Fan (✉) · C. Rue
International Food Policy Institute (IFPRI), Washington, DC, USA
e-mail: s.fan@cau.edu.cn

S. Fan
China Agricultural University, College of Economics and Management, Beijing, China

Smallholders face a mix of interrelated risks and challenges which threaten their livelihoods, food security and nutrition. Traditionally, the literature on smallholders has focused on challenges to their livelihood strategies, such as lack of human capital and limited access to infrastructure, markets and technologies. But smallholders have also become increasingly vulnerable to a spectrum of emerging climatic, health, price, and financial risks and challenges. Not only does the occurrence of these shocks endanger already fragile food production systems, but the mere likelihood of their occurrence makes some smallholders more risk-averse and likely to pursue more subsistence-oriented activities, thus causing smallholder poverty to persist (Dercon 2009).

The role of smallholder farms in advancing global food security and nutrition, as well as overall development, is increasingly seen in a broader context. The old wisdom that small is always beautiful because of efficiency gains can no longer be universally applied. Smallholders are not a homogeneous group that should be supported at all costs, but rather a diverse set of households living in different types of economies. Research suggests that small is still beautiful in countries where non-farm growth is weak and the rural population is increasing (such as in agriculture-based economies), but bigger is better where non-farm sectors are booming and the urban population is increasing (as in transforming and transformed economies) (Fan and Chan-Kang 2005). Thus, optimal farm size is a dynamic concept that changes as a country's overall economy grows and as non-agricultural sectors develop (Fan et al. 2013).

Small farmers can therefore prosper either through a 'move up' or a 'move out' strategy. While some small farmers have the potential to undertake profitable commercial activities in the agricultural sector and expand their farm operation, others should be supported in exiting agriculture and seeking non-farm employment opportunities.

More broadly, smallholders have a unique role to play in the new global development agenda—the SDGs which world leaders agreed upon in September 2015. Smallholder agriculture, especially if well-integrated into a diversified rural economy and agrifood value chains, can contribute even more to inclusive growth and employment generation. Even very poor subsistence farmers can be empowered to manage resources sustainably, and benefit from goals around education, peace and gender equality. Assistance, through measures such as safety nets and support through off-farm employment to diversify livelihoods, can also help develop rural communities and interrupt cycles of poverty, hunger and undernutrition. In addition to promoting more inclusive patterns of growth, this support can also cushion the short-term impact of transitioning to non-farm activities.

Although smallholder agriculture is often recognised as a vital sector for development, it has rarely enjoyed the policy and institutional support necessary to allow smallholders and rural economies to thrive. A commitment to treat smallholder farms as viable businesses is a key to unlocking the sector's potential to contribute to a broader development agenda. Enhancing the viability of smallholder farming could serve to both reduce rural poverty and improve food security and nutrition and contribute to the achievement of multiple SDGs.

2 The Important but Shifting Role of Smallholders

Ideas about the role of smallholders have evolved over time, and this role is increasingly being seen in a broader economic context. The discussion about smallholder farms should be expanded beyond a strict focus on small versus large farms, to reflect the idea that optimal farm size is a dynamic concept that changes as a country's overall economy grows and as non-agricultural sectors develop. Within this framework, interventions must be tailored to the different types of smallholder farms and the specific contexts in which they operate.

The backdrop to the debate on small versus large farms is the dominance of smallholder farming systems in the developing world. Worldwide, about half a billion farms are smaller than two hectares, and these farms are getting smaller in many countries (Hazell et al. 2007). The continuing decline is due to factors such as growing rural population, urban growth that is not labour-intensive, formal and informal barriers to rural–urban migration and distortionary land policies. Small farms are estimated to produce four-fifths of the developing world's food (FAO 2011a). Moreover, they are home to approximately two-thirds of the world's three billion rural residents, the majority of people living in absolute poverty and half of the world's undernourished people (IFPRI 2005). To gain a better understanding of the role that smallholders play in a country's development, it is important to first look at the broader context of agricultural development. Growth in agriculture has been shown to be an important part of the initial stage of transformation in many countries.

Agricultural growth can provide the economy with much-needed stimuli such as capital, labour and foreign exchange for finance and can fuel growth in non-agricultural sectors (see, e.g., de Janvry and Sadoulet 2009). The connection is not automatic, however, and varies according to country-specific circumstances, especially the country's potential for agricultural and non-agricultural (including minerals and manufacturing) sources of growth (Hazell et al. 2010). Past successes in promoting agricultural development, such as the Green Revolution in Asia, were grounded in interventions and reforms that supported equitable agricultural growth and were led by small farms (Hazell 2009). Policies that enabled smallholder participation in the Green Revolution included the equitable distribution of land and secure ownership and tenancy rights, alongside scale-neutral technologies, temporary input subsidies and large investments in infrastructure (such as roads and irrigation).

A large body of empirical research argues that there are efficiency benefits to smallholder farms. Studies have shown a strong inverse relationship between farm size and land productivity, with smaller farms generating higher per-unit farm output than larger farms (for a summary, see Heltberg 1998). The standard explanations for this inverse relationship focus on small farms' more intensive use of inputs and the lower costs associated with supervising family labour on small farms compared with hired labour on larger farms. Multiple studies, however, have called into question the absolute efficiency advantage of small farms (Helfand and Levine 2004; Barrett et al. 2010). These researchers have argued that larger commercial farms have an advantage in terms of finance, technology and logistics, and that the inverse relationship disappears above a certain farm size, or after factors such as land quality are taken into account—but even these studies have been challenged. A more dynamic argument about efficient farm size is that small farms have an advantage over large farms in terms of labour supervision and local knowledge, but larger farms gain the advantage as an economy shifts towards technologically advanced, capital-intensive, and market-oriented agriculture (Poulton et al. 2010).

One of the fundamental models of development economics asserts that the development of a dual-sector economy occurs through the transfer of low-productivity agricultural labour to the higher-productivity industrial and service sectors. The flow of labour continues until the marginal productivity of labour—in other words, income—is equal between the farm and non-farm sectors, after adjusting for labour quality and cost of living. This essentially means that workers will move from one sector to the other until wages are equal in the two sectors. Within this framework, farm size is an endogenous variable whose optimal value is the point of equal marginal productivity (again, income). Generally, it is expected that as labourers migrate out of rural areas, operational farm size will increase as those leaving agriculture sell or rent their land to the remaining farmers who can more efficiently expand their operations.

Yet, over the past few decades, farm structures in many developing countries have been affected by government policies that distort incentives for, and limit the extent of, efficiency-enhancing land transactions. (This is not to deny any justification for equity-oriented redistributive land reforms in certain highly unequal socioeconomic contexts.) Such interventions have included the imposition of ceilings on landholding size in a number of Asian countries, such as Bangladesh, India, Pakistan and the Philippines. Alternatively, many land-abundant developing countries, especially in SSA, have artificially promoted large-scale, commercial farms. These countries include post-independence Nigeria, Sudan and Tanzania, as well as the Democratic Republic of Congo and Mozambique, where more recent large land acquisition deals have taken place.

This artificial promotion of small or large farms, through restrictions on minimum or maximum landownership or rental, has been shown to result in inefficiencies by reducing farm productivity. For example, preliminary findings from the Philippines show that imposing a ceiling on farm size results in the misallocation of resources, causing agricultural labour productivity to drop by 7% and the share of employment in agriculture to increase from 45.1 to 48.5% (Adamopoulos and Restuccia 2013). The same can be seen in India and China, where reduced restrictions on land rental markets improved agricultural productivity by transferring land to more efficient (but often still poor) producers (Deininger and Jin 2005; Deininger et al. 2008). In fact, evidence from China shows that removing constraints on land rental markets has a much more positive impact on productivity gains and rental market participation than does administratively reallocating land, because the latter is weighed down by high transaction costs and imperfect information. Moreover, in Ethiopia, evidence shows that land tenure reform that expanded land renting contributes to improved tenure security, food security and child nutrition, whereas restrictions on land renting may contribute to deeper rural poverty traps and food insecurity (Holden and Otsuka 2014).

3 Typology of Development Pathways for Smallholders

Given the pivotal and substantial presence of smallholders in many developing countries, policies that directly or indirectly affect smallholder farmers have significant effects on the social and economic trajectory of those countries. However, the appropriate livelihood strategies should not be treated as a single pathway but instead as a dynamic process that reflects the different types of smallholders and economies (Table 1). We have created a typology that reflects the diversity of possible livelihood strategies and development pathways for smallholder farmers. This typology distinguishes between (1) the profitability of smallholders within the agricultural sector (subsistence farmers without profit potential, subsistence farmers with profit potential and commercialised smallholder farmers) and (2) the different stages of economic transformation (agriculture-based, transforming and transformed economies).

First, smallholders are a diverse set of households and individuals who face various constraints on their ability to undertake potentially profitable activities in the agricultural sector. Past studies have divided smallholders based on socioeconomic and biophysical variables such as population density, agricultural potential (determined by agro-ecological conditions such as water supply, soil fertility and biotic pressures from pests and diseases) and market access (Omamo et al. 2006). Other determinants of smallholder livelihood strategies include the asset position of households and the characteristics of the production environment (including institutions, power structures and market policies).

Within this typology, subsistence farmers are smallholders who consume the majority of their farm output and who are held back from participating more actively in commercially oriented agriculture by a variety of constraints. The potential to turn production systems into profitable enterprises is greatest among the subsistence farmers who are facing soft constraints—such as limited financial and human capital and asymmetric access to markets and information—that can be addressed through various policy and programmatic channels. In addition to soft constraints, the presence of hard constraints—such as marginal lands that are far from markets are limited in size and have extremely low rainfall and soil quality—severely hampers the ability of other smallholders to increase their production capacity and move towards profitable farming systems. Commercial smallholders are already involved in profitable agricultural activities but are held back from scaling up their commercial activities by factors such as limited access to capital and risk-reducing tools.

Second, the appropriate development pathway for smallholder farmers also depends on the level of transformation within the country's economy. The transformation process involves increased productivity and commercialisation in agriculture, alongside economic diversification and growth. The exact duration and character of the transformation vary across developing countries, but it includes several fundamental changes in the structure of the economy: a declining share of agriculture in gross domestic product (GDP) and employment, increasing rural-urban migration, the rise of a modern industrial and service economy, and a demographic transition to lower birth and death rates (Timmer 1988). In this typology, agriculture-based

Table 1 Move up or move out. *Source* adapted from Fan et al. (2014) and Fan et al. (2015)

WHICH PATH?	Whether a small Farmer should be targeted to 'move up' or 'move out' of agricultura depends on whether they fase the hard constraints that inhibit profit potential:	
SOFT CONSTRAINTS	**MOVE UP**	**MOVE OUT**
Limited access to markets and information	•	•
Limited financial capital	•	•
Limited access to infrastructure	•	•
Limited access to smallholder-friendly technologies	•	•
HARD CONSTRAINTS		
High population density		•
Low quality soil		•
Low rainfall and high temperaturas		•
Remote location		•
WHICH STRATEGY?	The best supportive strategies to aid farmers in either moving up or moving out depend on the type of economy:	
AGRICULTURE-BASED ECONOMY	**MOVE UP**	**MOVE OUT**
Productive cross-sector social safety nets that combine long-term tools with short-term support	•	
Investment in infrastructure, agricultural research and extensión, and smallholder-friendly and climate-smarttechnologies	•	
Access to innovative financial services	•	
Social safety nets		•
Nutrition-focused crop production for own consumption		•
Education and training for nonfarm employment		•
Migration to urban centers and other agricultura areas with greater profit potential		•
TRANSFORMING ECONOMY		
Flexible arrengements for land transfer	•	•
Risk reduction and management tools	•	
Access to market information	•	
Pro-smaller, nutrition-sensitive value chains	•	
Social safety nets	•	•
Improved access to housing, education, and health services for rural migrants	•	•
Vertical and horizontal coordination to meet safety, quality, and quantity standards	•	
Enhanced role of farmers' organizations, particularly for women farmers	•	
Education and training for nonfarm employment		•
TRANSFORMED ECONOMY		
Provide incentives for high-value production	•	
Reduced trade restrictions and subsidies	•	
Flexible arrengements for land transfer	•	•
Efficiency and quality-enhanced production systems	•	
Vertical and horizontal market coordination	•	
Social safety nets		•
Improved access to housing, education, and health services for rural migrants		•
Education and training for nonfarm employment		•

economies are those that derive a significant portion of their economic output and growth from the agricultural sector. This group includes most countries in SSA. Transforming economies, which lie mainly in East and South Asia, are those in which agriculture's significant role is being gradually replaced by the manufacturing and service sectors, although poverty continues to be heavily concentrated in rural areas. Finally, transformed countries, which are mainly in Eastern Europe and Latin America, are those in which agriculture has become a minor source of economic growth.

4 A Spectrum of Challenges Hinders the Profitability of Smallholder Farms

Smallholder farms are increasingly faced with a mix of challenges, including those that are naturally occurring and those that are caused by man, that influence their capacity to increase production and move towards profitable farming systems. These challenges lead farmers to undertake lower-risk and lower-yielding agricultural activities that perpetuate a cycle of poverty, including those with little or no profit. Women on small farms—who account for on average 43% of the agricultural labour force in developing countries—are particularly disadvantaged in accessing productive resources, such as land, livestock, agricultural inputs, technology, markets, and extension and financial services (FAO 2011b). Yet women play a vital role in improving agricultural output, enhancing food security and nutrition in the household and promoting overall development. High production constraints also make agriculture unattractive to young people—the very ones who can bring energy, vitality and innovation into the agricultural labour force in many developing countries (Brooks et al. 2013).

4.1 Limited Farm Size

Over the past few decades, high population growth and inheritance-based land fragmentation have resulted in decreasing farm size and high population density in many Asian countries and parts of Africa (Eastwood et al. 2009; Thapa and Gaiha 2011). Recent trends indicate that SSA will continue to experience declining farm size, while Asia is showing signs of farm consolidation (Jayne et al. 2014; Masters et al. 2013; Otsuka and Place 2014). An analysis of the relationship between increasing rural population density and smallholder farming systems in Kenya shows that, in addition to declining farm size and incomes, increasing rural population density is associated with decreasing agricultural labour productivity after a certain population density threshold (Muyanga and Jayne 2014). This inverse relationship is potentially the result of unsustainable agricultural intensification (Drechsel et al. 2001).

4.2 Access to Financial Services

Many small farmers are excluded from productivity-enhancing financial services, such as loans and saving accounts, and are thus unable to secure much-needed capital and lack the buffer against adversity and shocks that financial services offer. An analysis of maize farmers in Ghana reveals that small farms face more credit constraints than large farms (Kuwornu et al. 2012). In rural areas, where the majority of smallholders reside, access to formal financial services is particularly limited (Demirguc-Kunt and Klapper 2012). Reasons for this include dispersed demand and the high cost of service in low-population areas; weak administrative capacity of rural banks; agriculture-specific risks such as variable weather patterns, pests and price fluctuations that affect whole communities; and lack of formally defined property and land-use rights to act as collateral for loans.

4.3 Climate Change

The growing incidence and intensity of extreme weather events increasingly threaten the global food system (Zseleczky and Yosef 2014). If business as usual continues and the world becomes 3–4 °C warmer by 2050, crop yields could decline by 15–20% (World Bank 2013). In some African countries, yields from rainfed agriculture could decrease by up to 50% by 2020, with small-scale farmers being hit the hardest (IPCC 2007). In Malawi, smallholder farmers have experienced greater economic losses during droughts than have large landholders, in part because smallholders grew more drought-sensitive crops (Pauw et al. 2010). Smallholder farms are particularly vulnerable to more frequent extreme weather events because of such factors as chronic food insecurity, lack of access to formal safety nets, and high reliance on climate-dependent agriculture, coupled with limited resources and capacity for mitigating and adapting to the effects of climate change (Harvey et al. 2014).

4.4 Price Spikes and Volatility

Recent food price volatility and spikes have affected both producers and poor consumers. The complex set of factors behind the recent food price crises in 2007–2008 and 2011—including diversion of crops for biofuel, extreme weather events, low grain stocks and panicky trade behaviours—is still present or has the potential to re-emerge. The magnitude and direction of the impact on smallholder farms depend on several variables, including whether input costs increase, whether the farmers are net buyers or sellers of food, farmer capacity to step up production and to bring the increased output to market, and off-farm income (Anríquez et al. 2013). Recent studies in Bangladesh and Malawi suggest that an increase in the price of staple crops (rice and maize) resulted in a higher welfare loss for small landholders compared with large landholders (Karfakis et al. 2011).

4.5 Access to Modern Markets

Profitable market access by smallholders is challenged by a multidimensional set of factors. The participation of smallholders in modern market channels has a positive effect on their income, but participation is determined by a mix of non-land assets, with varied findings on the role of farm size in determining participation. These non-land assets include rural infrastructure (such as road access and irrigation), membership in cooperatives, education, modern market participation of nearby farms and rural non-farm employment (Hernandez and Reardon 2012; Rao and Qaim 2011). Lack of information (regarding price, supply and demand, and quality standards) leads smallholder farmers to face higher prices from opportunistic middlemen and traders, as well as lower market participation (Omiti et al. 2009). Amid rapid economic growth, urbanisation and globalisation, food supply channels are becoming longer geographically but shorter in terms of participants (Reardon et al. 2009).

5 Smallholder Farms Need to Move Up or Move Out

As stakeholders continue to deliberate on action plans for supporting sustainable smallholder farms, it is important to recognise that there is no 'one-size-fits-all' policy. The appropriate development pathway and livelihood strategies for each smallholder farm should reflect its particular characteristics and the level of transformation within the country's economy. Public policy should support smallholder farms in either moving up to commercially oriented and profitable farming systems, or moving out of agriculture to seek non-farm employment opportunities.

In agriculture-based economies, it is important to focus on advancing policies that move up small farmers that have the potential to become profitable by increasing their productivity. In both transforming and transformed economies, it is equally imperative to help such small farmers move up by promoting high-value agriculture and improving links to urban and global markets. For smallholder farms that are already profitable, policies to help scale up commercial activities are essential. Smallholder farms without profit potential, however, will require humanitarian assistance in the short term, and viable exit strategies out of agriculture to engage in urban and non-farm economic activities in the long term. To move smallholder farms with profit potential towards greater prosperity while at the same time improving global food security and nutrition and health outcomes, a number of steps must be taken, as outlined below.

5.1 Promote Land Rights and Efficient Land Markets

Given the heterogeneous character of economic growth and structures across developing countries, optimal farm size depends heavily on context, including the stage and structure of a country's economic and demographic development. Because well-functioning land sale and rental markets can have a major impact on agricultural productivity, governments in developing countries should not implement policies that promote cookie-cutter farm structures (for both rental and owner-occupied farms), which can lead to misallocation of resources.

Institutional reforms are needed to facilitate the efficient transfer of land through the certification of land rights and through well-functioning and transparent land sales and rental markets. Lifting restrictions on minimum or maximum landownership or land rental markets and securing property rights improves agricultural productivity. It does so by encouraging the transfer of land from small and poor farmers who have less ability or willingness to undertake agricultural activities (but who stay in agriculture due to fears of unfair compensation for land transfers) to more efficient (but often still poor) producers with more interest and resources (Deininger et al. 2008).

5.2 Enhance Risk Management, Mitigation and Adaptation Strategies

Smallholder farms urgently need better access to risk management tools and strategies to increase their resilience to a spectrum of shocks, including weather and price shocks. Tools such as index-based insurance can help farmers take productivity-enhancing risks, although their commercial viability for a smallholder clientele is still being studied. In the face of volatile crop prices, collaboration is needed among the private sector, governments and donors, to design innovative and flexible market-based price stabilisation tools—such as hedging in futures markets—that are suitable for smallholder farms. These tools limit the risk exposure of producers, without the distortionary effects and high costs of current price support measures (such as input, output and consumer price subsidies).

Reducing risks associated with price volatility requires supportive macroeconomic policies. National governments should encourage transparent, fair and open global trade, by eliminating formal and informal export restrictions and refraining from imposing new ones. Although export bans may help to secure domestic food supplies, they tend to exacerbate global price hikes, thus hurting the poorest net buyers of food. Food prices have been increasingly linked to energy prices, because of the growing diversion of food crops towards biofuel production as energy prices increase.

In terms of climate-induced shocks, a pro-poor climate change policy that creates value for smallholder farms and integrates them into global carbon markets is essential, although a viable modality has not yet been developed (De Pinto et al. 2010). Investments in triple-win agricultural practices and technologies can be effective in raising smallholder productivity alongside climate change mitigation and adaptation strategies (Bryan et al. 2013).

5.3 *Support Efficient and Inclusive Food Value Chains*

Linking smallholder farms to modern agrifood value chains is critical for improving agricultural productivity, food security and nutrition. Overcoming barriers to accessing modern value chains requires institutional innovations for coordination among smallholder farms, including group lending and producer associations. Such mechanisms require strong institutional capacity, in a stable policy environment that promotes private-sector investments adapted to the needs of smallholder farms. Information and communication technologies also offer the opportunity to link smallholder farms to markets, by helping them to reduce transaction costs, increase their bargaining power and acquire real-time market information. Financial services (bundled with, e.g., insurance) and investments in rural infrastructure also need to be scaled up. By bundling financial and non-financial solutions (such as insurance and agricultural advisory services), an environment that allows for comprehensive risk management solutions can be created (Vargas Hill and Torero 2009).

Furthermore, participation by smallholder farms in modern value chains can be leveraged for better nutrition and health. Greater investments in the development of nutrient-rich crop varieties accessible to the poor, coupled with public information campaigns and pricing policies, can help to increase the availability and consumption of nutritious foods (Hawkes and Ruel 2012). Sound regulatory and monitoring systems along the entire chain can also help to ensure that agricultural intensification does not harm people's health through, for example, foodborne and waterborne diseases, occupational hazards and environmental damage.

5.4 *Close Gender Gaps and Develop Young Farmers*

Addressing the inequity in access to productive resources, services and markets for women farmers (who account for a large percentage of smallholder farmers) is not only a rights issue, but also an efficiency issue. Gender inequality also leads to inefficient allocation of resources, which in turn means reduced agricultural productivity and poor nutrition and health outcomes. Evidence from Nigeria and Uganda suggests that lower productivity persists in female-owned plots and female-headed households (Peterman et al. 2010). Closing the gender gap in agriculture has high returns that accrue to the entire society, not just women (Meinzen-Dick and Quisumbing 2013).

Developing youth participation in agriculture is also essential to realise agricultural growth, improve food security and nutrition and promote overall development. Interventions to increase the profitability of smallholder farms should be targeted at young farmers. Such steps would include better agricultural training, improved land rights and enhanced access to financial and non-financial services.

5.5 *Scale up Productive Cross-Sector Social Safety Nets*

Productive cross-sector social safety nets that combine long-term tools (to build productive and resilient livelihood strategies) with short-term social safety support (to provide a cushion against shocks) can be of great benefit to small farmers. Ethiopia, for example, has created the Productive Safety Net Programme (PSNP) and Other Food Security Programme (OFSP)/Household Assets Building Programme (HABP), which provide a portfolio of productivity-enhancing mechanisms. These programmes are targeted at food-insecure households, most of which engage in small-scale farming (Berhane et al. 2014), and they are designed to ensure a minimum level of food consumption, protect and build assets and assist households in boosting income generated from agricultural activities. Based on recent evidence, the creation of the PSNP reduced the length of the hungry season by one-third compared with households with no programme benefits. Households with access to both PSNP and OFSP/HABP had even greater reductions in their hungry season and increases in their livestock holdings.

6 Conclusion

World agriculture will need to undergo major changes if the demands of a growing and increasingly rich and urban population are to be met, against a background of increasing scarcity of natural resources and other emerging challenges. Smallholders are an important part of the development equation. However, smallholders are not a homogeneous group, and development policies should not treat them as such. Instead, the development pathways of smallholders consist of dynamic processes that vary according to the constraints they face and the stage of economic transformation. While some smallholder farmers have the potential to undertake profitable commercial activities in the agricultural sector, other farmers should be supported in exiting agriculture and seeking non-farm employment opportunities. For smallholder farmers with profit potential, agriculture is risky in the face of climate change, price shocks, limited financing options and inadequate access to healthy and nutritious food. Smallholders can successfully adapt their livelihood strategies to these challenges but need a supportive policy environment.

These policies and investments should focus on (1) promoting land rights and efficient land markets; (2) enhancing risk management, mitigation and adaptation strategies; (3) supporting efficient and inclusive food value chains; (4) closing gender gaps and developing young farmers; and (5) scaling up productive cross-sector social safety nets. As with all public investments, the costs of investments and programmes designed to improve smallholders' productivity need to be compared with the likely benefits in each country. Public funds have alternative uses, such as other investments within or outside agriculture. Moreover, in many circumstances, agricultural development requires addressing the obstacles faced by groups of agricultural producers other than smallholders.

This chapter has identified several areas in which further research could shed light on the opportunities for smallholder farmers with profit potential to move from subsistence to commercially oriented agricultural systems, as well as the challenges to their doing so. It is now time for governments in developed and developing countries, the research community, and private companies to focus their investments, innovations and policies on helping these smallholders manage risk, improve their resilience to shocks, and increase their access to finance and capital, while promoting future growth. All of these measures, adapted to each country's stage of economic development and transformation, will play a critical but varying role in bringing down barriers to profitable and efficient agricultural operations by smallholders.

References

Adamopoulos, T., & Restuccia, D. (2013). The size distribution of farms and international productivity differences. Working Paper 480. Unpublished, Department of Economics, University of Toronto.

Anríquez, G., Daidone, S., & Mane, E. (2013). Rising food prices and undernourishment: A cross-country inquiry. *Food Policy, 38,* 190–202.

Barrett, C., Bellemare, M., & Hou, J. (2010). Reconsidering conventional explanations of the inverse productivity-size relationship. *World Development, 38*(1), 88–97.

Berhane, G., Gilligan, D., Hoddinott, J., Kumar, N., & Seyoum Taffesse, A. (2014). Can social protection work in Africa? The impact of ethiopia's productive safety net programme. *Economic Development and Cultural Change, 63*(1), 1–26.

Brooks, K., Zorya, S., & Gautam, A. (2013). *"Employment in agriculture: Jobs for Africa's youth", in 2012 global food policy report.* Washington, DC: International Food Policy Research Institute.

Bryan, E., Ringler, C., Okoba, B., Koo, J., Herrero, M., & Silvestri, S. (2013). Can agriculture support climate change adaptation, greenhouse gas mitigation, and rural livelihoods? Insights from Kenya. *Climatic Change, 118,* 151–165.

de Janvry, A., & Sadoulet, E. (2009). Agricultural growth and poverty reduction: Additional evidence. *World Bank Research Observer, 25*(1), 1–20.

De Pinto, A., Magalhaes, M., & Ringler, C. (2010). *Potential of carbon markets for small farmers: A literature review, Discussion Paper 1004.* Washington, DC: International Food Policy Research Institute.

Deininger, K., & Jin, S. (2005). The potential of land rental markets in the process of economic development: Evidence from China. *Journal of Development Economics, 78*(1), 241–270.

Deininger, K., Jin, S., & Nagarajan, H. (2008). Efficiency and equity impacts of rural land rental restrictions: Evidence from India. *European Economic Review, 52*(5), 892–918.

Demirguc-Kunt, A., & Klapper, L. (2012). Measuring financial inclusion: The global findex database. Policy Research Working Paper 6025. Washington DC: World Bank.

Dercon, S. (2009). Risk, poverty, and insurance. In R. Vargas Hill & M. Torero (Eds.), *Brief 3 in innovations in insuring the poor*. 2020 Vision Focus 17. Washington, DC: International Food Policy Research Institute.

Drechsel, P., Gyiele, L., Kunze, D., & Cofie, O. (2001). Population density, soil nutrient depletion, and economic growth in Sub-Saharan Africa. *Ecological Economics, 38*(2), 251–258.

Eastwood, R., Lipton, M., & Newell, A. (2009). Farm size. In P. Pingali & R. Evenson (Eds.), *Handbook of agricultural economics* (Vol. 4). Amsterdam: Elsevier.

Fan, S., & Chan-Kang, C. (2005). *Is small beautiful? Farm size, productivity, and poverty in Asian agriculture*. Agricultural Economics 32, s1: 135–146; World Bank, World Development Report 2008: Agriculture for Development, (Washington, DC, 2007).

Fan, S., Brzeska, J., Keyzer, M., & Halsema, A. (2013). From subsistence to profit: Transforming smallholder farms. Food policy report. Washington, D.C.: International Food Policy Research Institute (IFPRI). http://dx.doi.org/10.2499/9780896295582.

Fan, S., Brzeska, J., Keyser, M., & Halsema, A. (2014). *From subsistence to profit*. Washington DC: IFPRI Food Policy Report.

Fan, S., Brzeska, J., & Olofinbiyi, T. (2015). The business imperative: Helping small family farmers to move up or move out. In 2014–2015 Global food policy report. Chapter 4 pp. 25–31. Washington, D.C.: International Food Policy Research Institute (IFPRI). http://ebrary.ifpri.org/cdm/ref/collection/p15738coll2/id/129075.

FAO (Food and Agriculture Organization of the United Nations). (2011a). Save and Grow: A—olicymaker's guide to the sustainable intensification of smallholder crop production. Rome.

FAO (Food and Agriculture Organization of the United Nations). (2011b). The state of food and agriculture 2010–11—Women in agriculture: Closing the gender gap for development. Rome.

Harvey, C., Rakotobe, Z., Rao, N., Dave, R., Razafimahatratra, H., Rabarijohn, R, Rajaofara, H., MacKinnon, J. (2014). *Extreme vulnerability of smallholder farmers to agricultural risks and climate change in Madagascar*. Philosophical transactions of the royal society B 369, no. 1639.

Hawkes, C., & Ruel, M. T. (2012). Value chains for nutrition. In S. Fan & R. Pandya-Lorch (Eds.), *Reshaping agriculture for nutrition and health*. Washington, DC: International Food Policy Research Institute.

Hazell, P. (2009). Transforming agriculture: The green revolution in Asia. In D. Spielman & R. Pandya-Lorch (Eds.), *Millions fed: Proven successes in agricultural development* (pp. 25–32). Washington, DC: International Food Policy Research Institute.

Hazell, P., Poulton, C., Wiggins, S., Dorward, A. (2007). The future of small farms for poverty reduction and growth. 2020 Discussion Paper 42. Washington, DC: International Food Policy Research Institute.

Hazell, P., Poulton, C., Wiggins, S., & Dorward, A. (2010). The future of small farms: Trajectories and policy priorities. *World Development, 38*(10), 1349–1361.

Helfand, S., & Levine, E. (2004). Farm size and the determinants of productive efficiency in the Brazilian Centre-West. *Agricultural Economics, 31*(2–3), 241–249.

Heltberg, R. (1998). Rural market imperfections and the farm size-productivity relationship: Evidence from Pakistan. *World Development, 26*(10), 1807–1826.

Holden, S., & Otsuka, K. (2014). The roles of land tenure reforms and land markets in the context of population growth and land use intensification in Africa. *Food Policy*, 48, October, 88–97.

IFPRI (International Food Policy Research Institute). (2005). *The future of small farms: Proceedings of a Research Workshop*, Wye, UK, June 26–29, 2005. Washington, DC.

IPCC (Intergovernmental Panel on Climate Change). (2007). Climate change 2007: Synthesis report, contribution of working groups I, II and III to the fourth assessment report of the intergovernmental panel on climate change. Geneva.

Jayne, T. S., Chamberlin, J., & Headey, D. (2014). Land pressures, the evolution of farming systems, and development strategies in Africa: A synthesis, food policy, 48, October, 1–17, ISSN 0306-9192, http://dx.doi.org/10.1016/j.foodpol.2014.05.014.

Karfakis, P., Velazco, J., Moreno, E., Covarrubias, K. (2011). Impact of increasing prices of agricultural commodities on poverty, ESA working paper 11–14. Rome: FAO, Agricultural Development Economics Division.

Kuwornu, J., Ohene-Ntow, I., & Asuming-Brempong, S. (2012). Agricultural credit allocation and constraint: Analyses of selected maize farmers in Ghana. *British Journal of Economics, Management & Trade, 2*(4), 353–374.

Lowder, S., Skoet, J., Singh, S. (2014). What do we really know about the number and distribution of farms and family farms in the world? Background paper for the state of food and agriculture 2014, ESA working paper 14–02 (Rome: Food and Agriculture Organization of the United Nations (FAO), Agricultural Development Economics Division, 2014).

Masters, W., Andersson Djurfeldt, A., De Haan, C., Hazell, P., Jayne, T., Jirstrom, M., et al. (2013). Urbanization and farm size in Asia and Africa: Implications for food security and agricultural research. *Global Food Security, 2,* 156–165.

Meinzen-Dick, R., & Quisumbing, A. (2013). "Women in agriculture: Closing the gender gap," in 2012 global food policy report Chapter 4. pp. 37–47. Washington, D.C.: International Food Policy Research Institute.

Muyanga, M., & Jayne, T. (2014). Effects of rising rural population density on smallholder agriculture in Kenya. *Food Policy, 48,* 98–113.

Omamo, S., Diao, X, Wood, S., Chamberlin, J., You, L., Benin, S., Wood-Sichra, U., & Tatwangire, A. (2006). *Strategic priorities for agricultural development in Eastern and Central Africa* (Research report 150). Washington, DC: International Food Policy Research Institute.

Omiti, J., Otieno, D., Nyanamba, T., & McCullough, E. (2009). Factors influencing the intensity of market participation by smallholder farmers: A case study of rural and peri-urban areas of Kenya. *African Journal of Agricultural and Resource Economics, 3*(1), 57–82.

Otsuka, K., & Place, F. (2014). Changes in land tenure and agricultural intensification in Sub-Saharan Africa, WIDER working paper 051. Helsinki: United Nations University, World Institute for Development Economics Research.

Pauw, K., Thurlow, J., & van Seventer, D. (2010). *Droughts and floods in Malawi: Assessing the economywide effects, IFPRI discussion paper 962.* Washington, DC: International Food Policy Research Institute.

Peterman, A., Quisumbing, A., Behrman, J., & Nkonya, E. (2010). *Understanding gender differences in agricultural productivity in Uganda and Nigeria, discussion paper 1003.* Washington, DC: International Food Policy Research Institute.

Poulton, C., Dorward, A., & Kydd, J. (2010). The future of small farms: New directions for services, institutions, and intermediation. *World Development, 38*(10), 1413–1428.

Rao, E., & Qaim, M. (2011). Supermarkets, farm household income, and poverty: Insights from Kenya. *World Development, 39,* 784–796.

Hernandez R., & Reardon, T. (2012). Tomato farmers and modern markets in Nicaragua: A duration analysis. Selected paper prepared for presentation at the 2012 agricultural & applied economics association's annual meeting, Seattle, Washington.

Reardon, T., Barrett, C. B., Berdegué, J. A., & Swinnen, J. F. M. (2009). Agrifood industry transformation and small farmers in developing countries. *World Development, 37*(11), 1717–1727.

Thapa, G., & Gaiha, R. (2011). Smallholder farming in Asia and the Pacific: Challenges and opportunities. *Paper Presented at the International Fund for Agricultural Development Conference on "New Directions for Smallholder Agriculture,"* Rome.

Timmer, C. P. (1988). The agricultural transformation. In H. B. Chenery & T. N. Srinivasan (Eds.), *Handbook of development economics* (Vol. 1, pp. 275–332). Amsterdam: North-Holland.

Vargas Hill, R., & Torero, M. (2009). *Innovations in insuring the poor, 2020 Focus Brief 17.* Washington, DC: International Food Policy Research Institute.

World Bank. (2013). Turn down the heat: Climate extremes, regional impacts, and the case for resilience. A report for the World Bank by the Potsdam Institute for climate impact research and climate analytics. Washington, DC.

Zseleczky, L. & Yosef, S. (2014). Are shocks really increasing? A selective review of the global frequency, severity, scope, and impact of five types of shocks. *Conference Paper 5, 2020 Resilience*, Washington, DC: International Food Policy Research Institute.

6

The Superior Role of Agricultural Growth in Reducing Child Stunting: An Instrumental Variables Approach

Sébastien Mary and Kelsey Shaw

Abstract This chapter examines the impacts of agricultural growth and non-agricultural growth on the prevalence of child stunting in developing countries between 1984 and 2014. We find that a 10% increase in agricultural gross domestic product (GDP) per capita would reduce stunting by 2.9%, whereas a similar relative increase in non-agricultural GDP per capita would reduce stunting by only 2.2%. We confirm that agricultural growth is superior to non-agricultural growth in reducing child stunting. However, given the moderate amplitude of the estimated effects, it is unlikely that a pro-poor growth strategy, even one focussed on agriculture, would generate sufficient stunting reductions in line with the Sustainable Development Goals (SDGs). Policymakers may consider prioritising their efforts towards complementary direct nutritional investments. We also estimate the reverse causal impacts of stunting on sectoral growth. Stunting costs on average approximately 13.6% of potential non-agricultural GDP per capita and 3.4% of potential agricultural GDP per capita.

1 Introduction

A majority of the literature has focussed on the impacts of aggregate economic growth on child stunting. While the amplitude of the aggregate impacts continues to be a topic of debate (e.g. Heltberg 2009; Ruel et al. 2013; Harttgen et al. 2013; Headey 2013; Vollmer et al. 2014; Alderman et al. 2014; Smith and Haddad 2015; O'Connell and Smith 2016), the literature has failed to address a more fundamental and practical question; that is, whether agricultural growth is more effective at reducing stunting

S. Mary (✉)
Department of Economics, Driehaus College of Business, DePaul University, Chicago, USA
e-mail: smary@depaul.edu

K. Shaw
Coherent Economics, Chicago 60602, USA
e-mail: kelsey.shaw@coherentecon.com

than non-agricultural growth. This chapter aims to answer this question while also addressing the central role of agriculture in pro-poor growth strategies in reducing hunger.

Agricultural growth has been considered more effective than non-agricultural growth in reducing undernutrition, because of the existence of large multiplier effects and intersectoral linkages that result in higher labour demand and higher wages in rural areas, allowing households to consume more calories and diversify their diets (Johnston and Mellor 1961; World Bank 2007; FAO et al. 2012). The supposedly superior role of agricultural growth also relies on the fact that many poor households and children affected by stunting live in rural areas. If economic growth emanates from agriculture more so than other sectors, greater participation by poor people in agriculture could produce a much greater benefit to rural households, and as a result to children (Mellor 1976). There are, however, a few mediating factors—such as market concentration, output tradability or overuse of fertilisers—that could limit the ability of agricultural growth to reduce undernutrition (e.g. Collier and Dercon 2009; World Bank 2007; Brainerd and Menon 2014).

Few studies have analysed the impacts of sectoral growth on child stunting. For example, Webb and Block (2012) find that stunting responds substantially to agricultural growth, but less so to non-agricultural growth. Using a sample of 29 countries, they claim that growth originating from agriculture may be at least twice as effective as growth emanating from non-agriculture. Headey (2013) finds that while non-agricultural growth reduces child stunting, agricultural growth is found to have no (statistically significant) impact on child stunting. Additionally, he finds that the difference between the estimated impacts of agricultural and non-agricultural growths is not statistically significant. On the contrary, Mary et al. (2018a, b) find that the estimated impacts are relatively large, in that a 10% increase in agricultural (non-agricultural) GDP per capita reduces child stunting by 9.6% (8.4%), concluding that agricultural growth is superior to non-agricultural growth for generating child stunting reductions. Given the lack of consensus and research, whether agricultural growth is superior for reducing child stunting remains an open question.

Previous studies in the literature have also often ignored the existence of reverse causality between economic growth and stunting. Not only does growth affect stunting, but stunting may affect growth. This is important to consider, because ordinary least squares (OLS) estimates are biased downward, since the reverse causality is presumably negative, and therefore the impact of economic growth is overstated. Yet, some studies have attempted to account for the reverse causality. For instance, Webb and Block (2012) account for the endogeneity bias by using generalised method of moments (GMM) estimations. However, such estimators have been heavily criticised for their lack of reliability (Bazzi and Clemens 2013). Mary et al. (2018a, b) use a natural experiment based on temperature variations, but their approach relies on a strong identifying assumption: that temperature shocks have no physiological effects on child nutrition. Other studies have used instrumental variables, using cereal yields or investment rates as instruments (Vollmer et al. 2014; Smith and Haddad 2015), but the latter may not be plausibly exogenous instruments (Mary et al. 2018a, b; Mary 2018).

Instead, this chapter uses a novel approach to accounting for the reverse causal effect of stunting on agricultural and non-agricultural growths. The estimation strategy identifies the causal impact of agricultural and non-agricultural per capita GDPs (in logs) on stunting, by extending a two-step procedure that has been used in comparable contexts where the search for valid instruments has been especially elusive (e.g. Brückner 2013; Brückner and Lederman 2015). First, we estimate the reverse causal impacts of child stunting on both agricultural and non-agricultural per capita GDPs, using rainfall and temperature anomalies as instrumental variables (IVs) to generate exogenous variations in child stunting. In a second step, we estimate the effect that agricultural and non-agricultural growths have on stunting, using the residual agricultural and non-agricultural per capita GDPs that are not driven by stunting as an instrument. Not only does the approach estimate the impact of sectoral growth on stunting, it also estimates the reverse causal impacts of stunting on sectoral growth.

Using a dataset of 69 developing countries, between 1984 and 2014, we find that a 10% increase in agricultural (non-agricultural) GDP per capita would reduce stunting prevalence by 2.9% (2.2%). The impacts of agricultural growth are higher than the impacts of non-agricultural growth, but remain relatively modest. Still, we find statistical support that agricultural growth is more effective than non-agricultural growth. However, given the moderate amplitude of the estimated effects, it is unlikely that a pro-poor growth strategy, even focussed on agriculture, would generate sufficient stunting reductions in line with the SDGs. We also estimate the reverse causal impacts of stunting on current growth and find that a 1% point increase in stunting prevalence results in a 0.1% (0.4%) decrease in agricultural (non-agricultural) GDP per capita. A back-of-the-envelope calculation suggests that stunting costs approximately 13.6% of potential non-agricultural GDP per capita and 3.4% of potential agricultural GDP per capita. The remainder of the chapter is as follows. Section 2 describes the empirical model and the identification strategy. Section 3 presents the dataset. Section 4 analyses the results. Section 5 concludes.

2 Model and Identification Strategy

There are several conceptual frameworks that identify and distinguish causal factors to help comprehend undernutrition. For example, the UNICEF model (1998) classifies the determinants of nutrition into three categories: immediate, underlying and basic. This classification has often provided a theoretical background for modelling stunting, and most empirical studies have typically focussed on one category. The scope of this chapter covers the most structural factors, or basic determinants, with a focus on sectoral economic growth. In line with most studies in the literature, we do not include control variables from a different group of causal factors.

2.1 *Empirical Model*

The model can be expressed as follows:

$$y_{it} = \alpha x_{it}^{\text{AG}} + \beta x_{it}^{\text{NONAG}} + \gamma z_{it} + c_i + d_t + \varepsilon_{it} \quad (1)$$

where country is indexed by i and year is indexed by t; y_{it} is the prevalence of child stunting of country i in year t; x_{it}^{AG} is the weighted logarithm of agricultural GDP per capita; x_{it}^{NONAG} is the weighted logarithm of non-agricultural GDP per capita; z_{it} is a vector of independent variables, including a composite index of governance and urbanisation. c_i is a country-specific effect. d_t represents time dummies. The inclusion of country and period fixed effects accounts for the presence of time-invariant omitted variables and common shocks affecting both stunting and sectoral growth, respectively.

Separating GDP into sectoral components allows us to investigate their relative roles in reducing stunting. Following the existing literature, each sector's logged GDP is weighted by its share in total GDP, to account for the size of each sector at the country level. This accounts for the fact that the impact of a given increase in agricultural GDP per capita, or agricultural growth, is likely to be small (large) in a country with a small (large) agricultural sector such as Chile (Cameroon). Also, the introduction of sectoral weights implicitly accounts for the country's (more or less advanced) stage of development.

In line with the existing literature, we include several other independent variables, namely urbanisation and governance.[1] The inclusion of these variables is guided by the desire to avoid the loss of stunting observations, as well as to follow the existing literature (e.g. Smith and Haddad 2015; Mary et al. 2018a, b). We also test a more extended version of this model in robustness analyses.

Given our focus on nutrition, we follow Smith and Haddad (2015) and define governance as the set of traditions, policies and institutions (and implicitly their effectiveness) that work towards ensuring food security, especially for children. For example, more democratic governments are more likely to ensure that the benefits of higher national income reach those who need it the most, by spending a higher share of their revenues on social and nutrition-relevant infrastructures. Also, the urbanisation rate controls for the increasing migration of populations towards urban areas in many developing countries, where the provision of nutrition and social services is of higher quality and likely more stable.

Mary et al. (2018a, b) argue that a stunting prevalence observation in a specific year, for example 2000, is affected by factors determined as early as 1995, and recommend using a model where regressors are replaced by their five-year moving averages. Equation (1) can therefore be rewritten as:

$$y_{it} = \delta x_{it}^{\text{AG-MA5}} + \theta x_{it}^{\text{NONAG-MA5}} + \vartheta z_{it}^{MA5} + f_i + g_t +_{it} \quad (2)$$

---

[1] Most studies in the aggregate literature typically include one or no additional independent variables.

where the superscript MA5 denotes the five-year rolling average of the regressor. For example, the five-year moving average of agricultural growth is calculated as follows:

$$x_{it}^{\text{AG-MA5}} = \frac{(x_{it}^{\text{AG}} + x_{it-1}^{\text{AG}} + x_{it-2}^{\text{AG}} + x_{it-3}^{\text{AG}} + x_{it-4}^{\text{AG}})}{5}$$

2.2 *Estimation*

As explained in the introduction, finding a strategy to account for reverse causality can be challenging. In this chapter, we extend the two-step approach devised by Brückner (2013). As a first step, we estimate the effect that stunting has on both agricultural and non-agricultural per capita GDPs, using an instrumental variable (IV) approach.[2] Equations (3) and (4) present the first-step estimations[3]:

$$x_{it}^{\text{AG-MA5}} = \pi y_{it} + \rho x_{it}^{\text{NONAG-MA5}} + \sigma z_{it}^{\text{MA5}} + h_i + r_t + \tau_{it} \quad (3)$$

$$x_{it}^{(\text{NONAG-MA5})} = \varphi y_i t + \omega x_i t^{(\text{AG-MA5})} + \zeta z_i t^M A5 + j_i + k_t + \xi_{it} \quad (4)$$

The statistical significance of π and φ provides direct endogeneity tests of stunting to agricultural and non-agricultural growths, respectively. Estimating the effect that stunting has on agricultural and non-agricultural GDP per capita requires an exogenous source of variation for stunting. Therefore, rainfall and temperature anomalies, defined here as deviations from the long-run mean calculated over the full sample period—and higher moments—are used as excluded instruments in the two-stage least squares (2SLS) estimation of Eq. (2). Weather anomalies (z) are generally defined as follows:

$$z_{it} = \frac{a_{it} - \overline{a_i}}{\overline{a_i}}$$

where $\overline{a_i}$ is the mean temperature (rainfall) level, observed over the full-time period, in country i; a_{it} is the temperature (rainfall) level in year t. If z_{it} is positive (negative), this indicates that the average temperature (rainfall) in year t in country i is above (below) the long-run level. Higher moments of z capture potentially nonlinear effects.

Then, in a second step, if we construct adjusted per capita sectoral GDP series, where the responses of per capita agricultural and non-agricultural GDP to stunting have been 'partialled out' using estimates from Eqs. (3) and (4):

---

[2]We thank Markus Bruckner for the suggestion on how to extend his approach to two endogenous variables.

[3]We also account for intersectoral linkages by considering non-agricultural GDP per capita endogenous in Eq. (3) and agricultural GDP per capita endogenous in Eq. (4).

$$x_{it}^{\text{AG}*} = x_{it}^{\text{AG-MA5}} - \pi y_{it} - \rho x_{it}^{\text{NONAG-MA5}} - \sigma z_{it}^{\text{MA5}}$$

$$x_{it}^{\text{NONAG}*} = x_{it}^{\text{NONAG-MA5}} - \varphi y_{it} - \omega x_{it}^{\text{AG-MA5}} - \zeta z_{it}^{\text{MA5}}$$

$x_{it}^{\text{AG}*}$ and $x_{it}^{\text{NONAG}*}$ are free of the endogeneity bias and are, respectively, used to instrument $x_{it}^{\text{AG-MA5}}$ and $x_{it}^{\text{NONAG-MA5}}$ in estimating Eq. (2) via 2SLS. The second-step estimation is exactly identified. Instruments are also included as independent variables in the second-step estimation. Other independent variables are assumed exogenous.

2.3 *Identification*

What makes this identification strategy plausible? First, temperature or rainfall anomalies are, by definition, exogenous and random. Second, weather shocks affect the physiology of children and are therefore likely correlated with stunting prevalence (Millward 2017). Third, stunting is not fully determined within the first two years of a life, and as such is not a static phenomenon. Stunting has been portrayed as a chronic condition because of growth faltering (i.e. height-for-age of children observed later in life is, to some degree, a reflection of failure to grow before age 2). However, there is an intense ongoing debate as to whether substantial height catch-up can occur between 24 months and mid-childhood, even in the absence of any interventions (Prentice et al. 2013a, b; Prendergast and Humphrey 2014; Leroy et al. 2013). In a seminal paper, Prentice et al. (2013a, b) show that a pubertal growth phase in rural Gambian children allows very considerable height recovery and conclude that their data are sufficient to dispel the myth of life-long size entrainment by age 24 months (Prentice et al. 2013b).

The main assumption for our identification strategy is that current weather anomalies are physiological shocks that affect stunting, and hence sectoral GDP per capita. Given the high incidence of child labour in developing countries,[4] it is expected that the impacts will affect both agricultural and non-agricultural sectors, though it is not clear what type of growth will suffer most from the burden of child undernutrition.

What about the exclusion restriction for the estimation of Eqs. (3) and (4)? Given the modelling framework, the exclusion restrictions are valid by design. Indeed, it would be impossible for current weather anomalies to affect variables that now mainly represent previous sectoral growth. Additionally, the Hansen J-test for over-identification of all instruments is used to investigate the validity of our instruments. This tests the null hypothesis that the instruments are uncorrelated with the error term. If the p-value is below 10%, we can conclude that all instruments are not exogenous and therefore invalid. The strength of instruments is further examined by reporting the Kleibergen-Paap rk Wald F statistic (or F-statistic), which we compare with the

[4]It is estimated that more than one in five children in Africa are employed against their will in stone quarries, farms and mines (ILO 2017).

critical values from Stock and Yogo (2005) for testing weak instruments. We test the null hypothesis that the maximum relative bias and size distortions are greater than 10, 15 or 20%.

3 Data

In this study, we initially collect a dataset on child stunting and several basic determinants, for a sample of 86 low-income and middle-income countries between 1984 and 2014. Given that the impact of governance needs to be observed over the long run, the selection of countries and years is determined by available data on governance. The Political Risk Services (PRS) dataset from the International Country Risk Guide combines a large country coverage and long-time series on multiple dimensions of governance and has been used extensively in empirical research. We gather data for all low-income and middle-income countries and for the years available in the PRS dataset. We then exclude countries with less than two observations on child stunting over the period. We find no outlier using the Hadi procedure (based on 5% level). The final sample includes 69 countries for 369 observations.

In line with the most recent studies, we use stunting (height-for-age) as the dependent variable, rather than wasting or underweight (weight-for-age). In doing so, we acknowledge that stunting has become a reference variable because it better captures the process of undernutrition in the medium run.

Data for the prevalence of stunted children under age 5 are taken from the World Health Organization (WHO). Such data have been used in previous studies (Ruel et al. 2013; Smith and Haddad 2015; Mary et al. 2018a, b) and have the primary advantage of possessing a larger country coverage than the Demographic and Health Surveys. The prevalence of stunting is the percentage of children under age 5 whose height-for-age is more than two standard deviations below the median for the international reference population aged 0–59 months. For children up to 2 years old, height is measured by recumbent length. For older children, height is measured by stature while standing. The data is based on the child growth standards released in 2006 by the WHO. While there are other stunting variables that cover partial life range (24–36 months), most studies use the 0–59 months reference. Therefore, our decision to use this measure is driven by the need for comparison with existing studies. This measure has also often been used in policy discussions and has attracted much attention across all development stakeholders. It is also now directly recognised in the SDG agenda (SDG 2, Target 2.2).

For agricultural growth and non-agricultural growth, we use data from the World Bank database. GDP per capita is GDP divided by mid-year population. GDP is the sum of gross value added by all resident producers in the economy, plus any product taxes and minus any subsidies not included in the value of the products. Data are in constant USD 2011 (in purchasing power parity). The decomposition of GDP is based on the calculation of each sector's GDP (per capita) and the size of each sector within the economy; it is based on the industrial origin of value added, from the International Standard Industrial Classification (ISIC), Revision 3.

The PRS dataset provides 12 indicators, based on expert analyses, covering both political and social attributes to capture governance in its different characteristics. We use an index of governance similar to the one used in Smith and Haddad (2015), combining: (1) Bureaucratic Quality—the institutional strength and quality of the bureaucracy; (2) Law and Order—the strength and impartiality of the legal system and popular observance of the law; (3) Government Stability—the government's ability to carry out its declared programme and associated policies, and its ability to stay in office; (4) Corruption within the political system—a threat to foreign investment by distorting the economic and financial environment, reducing the efficiency of government and business by enabling people to assume positions of power through patronage rather than ability, and introducing inherent instability into the political process; and (5) Democratic Accountability—a measure of how responsive government is to its people, on the basis that the less responsive it is, the more likely it is that the government will fall—peacefully in a democratic society, but potentially violently in a non-democratic one (PRS 2016).

The index is the simple mean of these five indicators for each country-year data point, after placing each on a 0-1 scale to establish equivalent ranges and weights. Each dimension of governance, as shown above, can directly or indirectly promote reductions in child stunting. For example, bureaucratic effectiveness and political stability directly contribute to the quality and reliability of the provision of nutrition and social services that support many children's nutritional status. Similarly, democratic accountability ensures that pressures are maintained for public action because increased awareness of the issue of nutrition is likely to make governments more responsive to the needs of food insecure people, especially children.

Other variables are taken from the World Development Indicators (WDIs). Temperature data are from the Climate Research Unit (CRU) and the Tyndall Centre for Climate Change Research (TYN) of the University of East Anglia. Table 3 in Appendix displays descriptive statistics.

4 Results

This chapter examines whether agricultural and non-agricultural growths reduce the prevalence of child stunting. Constant terms are included in the estimations. We use country-clustered standard errors that are robust to heteroscedasticity.

4.1 *Effect of Child Stunting on Sectoral Growth*

Table 1 displays the estimates for the effect that stunting has on agricultural growth in column 2, and on non-agricultural growth in column 4. The associated first-stage

Table 1 Effect of child stunting prevalence on sectoral growth

Dependent variable	First stage	2SLS-IV	First stage	2SLS-IV
	(1)	(2)	(3)	(4)
	Stunting	AG growth	Stunting	NAG growth
Child stunting prevalence		**−0.001*****		**−0.004*****
		[0.00]		**[0.00]**
Urbanisation		0.016#		0.024
		(1.52)		(1.37)
Governance index		0.006		0.122
		(0.03)		(0.44)
Rainfall anomaly	−2.527		−2.527	
	(−1.33)		(−1.33)	
Temperature anomaly	−2.443***		−2.443***	
	(−5.71)		(−5.71)	
Temperature anomaly, squared	0.142**		0.142**	
	(2.14)		(2.14)	
Temperature anomaly, cubic	0.083***		0.083***	
	(5.41)		(5.41)	
Observations	369	369	369	369
Number of countries	69	69	69	69
Country fixed effects	Yes	Yes	Yes	Yes
Year fixed effects	Yes	Yes	Yes	Yes
Hansen J, *p*-value	n.a.	0.43	n.a.	0.38
First-stage *F*-stat	n.a.	11.01	n.a.	15.74
Stock–Yogo 10% size	n.a.	16.87	n.a.	16.87
Stock–Yogo 20% size	n.a.	7.54	n.a.	7.54
Stock–Yogo 10% relative bias	n.a.	7.56	n.a.	7.56
Stock–Yogo 20% relative bias	n.a.	5.57	n.a.	5.57

Source Authors' calculations
AG agricultural; *NAG* non-agricultural
Notes *, **, ***: significant at 10%, 5%, 1%. #: marginally insignificant ($p < 0.15$). *t*-statistics in parentheses. Each sector's growth is weighted by the sector's share of value added in total GDP

estimates that link rainfall and temperature anomalies to stunting are respectively displayed in columns 1 and 3. Increased rainfalls from their long-run levels are associated with decreased stunting, while temperature anomalies have various effects depending on the size of the anomaly. In other words, small-scale (large-scale) positive temperature anomalies decrease (increase) stunting.

The 2SLS estimate in column 2 of Table 1 suggests that a percentage point increase in child stunting prevalence decreases agricultural GDP per capita by 0.1%. A similar increase in child stunting would also decrease non-agricultural GDP per capita by 0.4% according to the estimate in column 4. A quick calculation based on the estimates in columns 2 and 4 suggests that, on average, stunting costs developing countries about 13.6% of potential non-agricultural GDP per capita and 3.4% of potential agricultural GDP per capita.

More fundamentally, the existence of negative reverse causal effects implies that OLS estimates of the impact of sectoral growth on stunting would be biased downwards. For both estimations in columns 2 and 4, the @@F-statistics suggest that weak instruments are not a problem as they are well above the 20% Stock–Yogo critical values. The p-values associated with the Hansen J-test are greater than 10% in columns 2 and 4, indicating that the tests do not reject that the instruments are uncorrelated with the second-stage error.

4.2 *Effect of Sectoral Growth on Child Stunting*

Table 2 shows the estimates of the effect of both agricultural and non-agricultural growths on stunting. The 2SLS coefficients are also higher than their OLS counterparts, in line with the existence of negative reverse causality. More importantly, the 2SLS coefficients (using the residuals calculated on the estimates in columns 2 and 4 in Table 1) are negative, significant at 1% in column 2, and imply that a 10% increase in agricultural (non-agricultural) GDP per capita would reduce stunting prevalence by 2.9%[5] (2.2%).

These findings are partially in line with Webb and Block (2012), who use a semi-parametric regression. However, we do not find supporting evidence that agricultural growth is twice as effective as non-agricultural growth, perhaps because we use a much larger dataset. Also, our results pertaining to the role of agricultural growth seem to be in contrast with Headey (2013). Despite this, the latter combines disaggregated state data for India (rather than country-level data) as well as country data in his dataset. Therefore, some caution must be applied when comparing our results with this study. Further, we find impacts that are much lower than in Mary et al. (2018a, b).

Furthermore, the 2SLS coefficient for agricultural growth is larger than for non-agricultural growth, suggesting the relative superiority of agricultural growth in the fight against children's food insecurity. A Wald test shows that the difference between agricultural growth and non-agricultural growth effects is indeed statistically significant, though at a 10% level (p-value: 0.05). That is, we find statistically significant support that agricultural growth is superior to non-agricultural growth towards reducing the number of stunted children in developing countries.

This result seems to be in contrast with Headey (2013), who finds no supporting evidence that the structure of growth matters for stunting reductions, but in line with Mary et al. (2018a, b). Also, we note that the OLS estimates in column 1 of Table 2 imply that the sectoral composition of growth does not affect child stunting

[5]This is calculated as: $\frac{9.848}{33.87} * \frac{10}{100} \approx 2.9\%$.

Table 2 Effect of sectoral growth on child stunting

Child stunting prevalence	OLS-FE	2SLS-IV
	(1)	(2)
Agricultural growth	**−10.249*****	**−9.848*****
	(−3.40)	**(−3.77)**
Non-agricultural growth	**−10.790*****	**−7.312*****
	(−4.64)	**(−3.26)**
Urbanisation rate	−0.472***	−0.343**
	(−4.03)	(−2.16)
Governance index	−17.982***	−20.489***
	(−3.33)	(−2.59)
Rainfall anomaly		−2.874#
		(−1.63)
Temperature anomaly		−1.999***
		(−4.61)
Temperature anomaly, squared		0.104*
		(1.71)
Temperature anomaly, cubic		0.071***
		(4.14)
Observations	369	369
Number of countries	69	69
Country fixed effects	Yes	Yes
Year fixed effects	Yes	Yes
Ramsey RESET test	0.04	0.13
First-stage F-stat	n.a.	648.9
H0: AG > Non-AG, p-value	0.64	0.05

Source Authors' calculations
Notes *, **, ***: significant at 10%, 5%, 1%. #: marginally insignificant ($p < 0.15$). t-statistics in parentheses. Each sector's growth is weighted by the sector's share of value added in total GDP. RESET tests are based on GMM distance

reductions. This may suggest that ignoring endogeneity may lead to wrong policy implications with respect to the relative importance of agricultural growth towards reducing child nutrition.

Lastly, we find that governance reduces stunting. Our estimate is negative, significant at 1% but somewhat higher than the estimate found in Smith and Haddad (2015). This suggests that governance plays a key role in food security. The coefficient for the urbanisation rate is also negative and significant at 5%. The Kleibergen-Paap statistic is well above the Stock–Yogo critical values and suggests that weak instruments are not a problem. Finally, the RESET test seems to indicate that an appropriate form has been used (p-value: 0.13), and omitted variable bias is likely not a problem.

4.3 Robustness Analyses

We investigate the robustness of our empirical results to the: (1) exclusion of year dummies; (2) use of data for sub-Saharan Africa (SSA) and Asia only (where the majority of stunted children live); (3) exclusion of Bangladesh, which is the country with the highest number of stunting observations in the sample; (4) use of a basic model; and (5) use of an extended model. The extended model includes trade openness (exports plus imports, divided by GDP), following Levine and Rothman (2006), and a measure of the military's involvement in politics from the PRS dataset. Trade may affect children's health through a number of pathways, including the degree to which governments are willing and able to fund public health, higher access to clean water, or health care (Levine and Rothman 2006). Military involvement may stem from an external or internal threat, be symptomatic of underlying difficulties, or be a full-scale military takeover. The military may also control food supply or use it as a weapon or payment for political support, therefore disrupting its distribution to the populations who need food the most.

Estimation tables can be found in Appendix Table 4. Our overall conclusion from these robustness tests is that the main pattern of results remains the same. We summarise the main results below:

(1) For all alternative specifications, excluding the one without year dummies, agricultural growth is statistically superior to non-agricultural growth in reducing child stunting (only marginally for the basic model).

(2) Focussing on SSA and Asia does not change our empirical results (column 2, Table 4).

(3) Dropping Bangladesh from our sample does not affect the empirical results (column 3, Table 4).

(4) Sectoral growth coefficients are much lower in the model without year dummies and the basic model (columns 1 and 4, Table 4).

(5) Non-agricultural growth is found to have no statistically significant impact on stunting in the basic model (column 4, Table 4).

(6) We find that trade openness decreases stunting, while a higher military presence in politics is associated with higher stunting in the extended model (column 5, Table 4).

5 Conclusions

This chapter has estimated and compared the impacts of increases in agricultural and non-agricultural GDP per capita on the prevalence of child stunting. We find that both types of growth reduce stunting. In particular, our findings show that a

10% increase in agricultural (non-agricultural) GDP per capita reduces child stunting, on average, by 2.9% (2.2%). We further examine whether the sectoral composition of growth matters for stunting reductions and find evidence that growth emanating from agriculture is more effective at reducing stunting than non-agricultural growth. This provides support for the view that the agricultural sector is the most important sector for fighting undernutrition.

However, given the relatively moderate amplitude of the estimated effects, it is unlikely that a pro-poor growth strategy, even focussed on agriculture, would generate sufficient stunting reductions in line with the SDGs. While Dercon (2013) questions the putative superior role of agriculture in development, our results would seem to suggest that economic growth, whether it originates from agriculture or not, may not have enough *oomph* to generate large reductions in child stunting. In fact, given the available evidence on nutrition-specific and nutrition-sensitive investments (e.g. Bhutta et al. 2008; Mary et al. 2018a, b), policymakers should consider prioritising their efforts towards complementary direct nutritional investments.

Acknowledgements The authors thank Sergio Gomez y Paloma and Laura Riesgo for useful comments. Data and Stata codes are available upon request.

Appendix

See Tables 3 and 4.

Table 3 Descriptive statistics of sample

Variables	(1)	(2)	(3)	(4)	(5)
	N	Mean	Standard deviation	Min	Max
Child stunting prevalence	369	33.87	14.55	4.3	76.7
Agricultural GDP per capita, weighted log	369	1.39	0.75	0.13	3.80
Non-agricultural GDP per capita, weighted log	369	6.09	1.60	2.39	9.16
Governance index	369	0.50	0.10	0.16	0.75
Urbanisation	369	42.29	18.15	10.26	86.46
Trade openness	369	64.61	32.39	13.04	222.25
Military in politics	369	0.49	0.24	0	1
Rainfall anomaly	369	0.005	0.13	−0.50	0.46
Temperature anomaly	369	0.011	0.59	−6.93	4.93

Table 4 The effect of sectoral growth on child stunting: robustness analyses

Child stunting prevalence	2SLS-IV	2SLS-IV	2SLS-IV	2SLS-IV	2SLS-IV
	(1)	(2)	(3)	(4)	(5)
	No year dummies	SSA and SEA	No Bangladesh	Basic model	Extended model
Agricultural growth	**−5.095****	**−13.516*****	**−9.865*****	**−5.095#**	**−14.325*****
	(−2.13)	**(−3.82)**	**(−3.67)**	**(−1.55)**	**(−4.68)**
Non-agricultural growth	**−4.778****	**−10.613*****	**−7.877*****	**−3.449**	**−8.398*****
	(−2.41)	**(−3.59)**	**(−3.33)**	**(−1.17)**	**(−4.14)**
Urbanisation rate	−0.767***	−0.301	−0.242*		−0.332*
	(−5.68)	(−1.32)	(−1.75)		(−1.90)
Governance index	−20.755***	−22.776**	−11.721***		−26.627***
	(−3.57)	(−2.18)	(−2.60)		(−2.90)
Trade openness					−0.092*
					(−1.85)
Military in politics					5.224*
					(1.91)
Coefficients for instruments omitted for the sake of space					
Observations	369	240	349	372	369
Number of countries	69	39	68	69	69
Ramsey RESET test	0.07	0.00	0.29	0.00	0.02
First−stage F−stat	977.5	466.5	1237	279.6	481.9
H0: AG > NAG, *p*-value	**0.41**	**0.04**	**0.09**	**0.14**	**0.00**

Source Authors' calculations
AG agricultural growth
Notes *, **, ***: significant at 10%, 5%, 1%. #: marginally insignificant ($p < 0.15$). *t*-statistics in parentheses. Each sector's growth is weighted by the sector's share of value added in total GDP. RESET tests are based on GMM distance

References

Alderman, H., Haddad, L., Headey, D., & Smith, L. (2014). Association between economic growth and early childhood nutrition. *The Lancet Global Health, 2*(9), e500. https://doi.org/10.1016/S2214-109X1470266-9.

Bazzi, S., & Clemens, M. A. (2013). Blunt instruments: Avoiding common pitfalls in identifying the causes of economic growth. *American Economic Journal: Macroeconomics, 5*(2), 152–186.

Bhutta, Z. A., Ahmed, T., Black, R. E., Cousens, S., Dewey, K., Giugliani, E., Haider, B. A., Kirkwood, B., Morris, S. S., Sachdev, H. P. S., Shekar, M. (2008). What works? Interventions for maternal and child undernutrition and survival. *The Lancet, 371*(9610), 417–440.

Brainerd, E., & Menon, N. (2014). Seasonal effects of water quality: The hidden costs of the Green Revolution to infant and child health in India. *Journal of Development Economics, 107,* 49–64. https://doi.org/10.1016/j.jdeveco.2013.11.004.

Brückner, M. 2013. On the simultaneity problem in the aid and growth debate. *Journal of Applied Econometrics,* 28(1):126–150.

Brückner, M., & Lederman, D. (2015). Trade openness and Economic growth: Panel data evidence from Sub-Sahran Africa. *Economica, 82,* 1302–1323.

Climatic Research Unit, University of East Anglia. http://www.cru.uea.ac.uk/, Dataset, Accessed February 1, 2016.

Collier, P., Dercon, S. (2009). *African Agriculture in 50 years: Smallholders in a rapidly changing world?* Paper Presented at the 2009 Food and Agriculture Organization Expert Meeting on "How to Feed the World in 2050". ftp://ftp.fao.org/docrep/fao/012/ak983e/ak983e00.pdf.

Dercon, S. (2013). Agriculture and development: revisiting the policy narratives. *Agricultural Economics, 44*(s1), 183–187. https://doi.org/10.1111/agec.12062.

FAO, WFP and IFAD. (2012). *The state of food insecurity in the world 2012. Economic growth is necessary but not sufficient to accelerate reduction of hunger and undernutrition.* Rome, FAO.

Harttgen, K., Klasen, S., & Vollmer, S. (2013). Economic growth and child undernutrition in sub-Saharan Africa. *Population and Development Review, 39,* 397–412. https://doi.org/10.1111/j.1728-4457.2013.00609.x.

Headey, D. (2013). Developmental drivers of nutritional change: A cross-country analysis. *World Development, 42,* 76–88. https://doi.org/10.1016/j.worlddev.2012.07.002.

Heltberg, R. (2009). Undernutrition, poverty and economic growth. *Health Economics, 18,* S77–S88. https://doi.org/10.1002/hec.1462.

ILO. (2017). Child labour in Africa. http://www.ilo.org/ipec/Regionsandcountries/Africa/lang–en/index.htm.

Johnston, B., & Mellor, J. (1961). The role of agricultural in economic development. *The American Economic Review, 4,* 566–593.

Leroy, J. L., Ruel, M., & Habicht, J. P. (2013). Critical windows for nutritional interventions against stunting. *American Journal of Clinical Nutrition, 98*(3), 854–855.

Levine, D. I., & Rothman, D. (2006). Does trade affect child health? *Journal of Health Economics, 25*(3), 538–554.

Mary, S. (2018). *How much does economic growth really contribute to child stunting reductions?* Working Paper, DePaul University.

Mary, S., Saravia-Matus, S., Gomez y Paloma, S. (2018a). Does nutrition-sensitive aid reduce the prevalence of undernourishment? *Food Policy, 74,* 100–116.

Mary, S., Shaw, K., Gomez y Paloma, S. (2018b). Does the sectoral composition of growth affect child stunting reductions? *Development Policy Review* (Accepted article).

Mellor, J. (1976). *The new economics of growth: A strategy for India and the developing world.* Ithaca, NY, USA: Cornell University Press.

Millward, D. J. (2017). Nutrition, infection and stunting: the roles of deficiencies of individual nutrients and foods, and of inflammation, as determinants of reduced linear growth of children. *Nutrition Research Reviews, 30,* 50–72.

O'Connell, S. A., Smith, C. (2016). Economic growth and child undernutrition. *The Lancet Global Health 4,* December, Correspondence.

Political Risk Services. (2016). *International country risk guide.* https://www.prsgroup.com, Dataset. Accessed February 1, 2016.

Prendergast, A. J., & Humphrey, J. H. (2014). The stunting syndrome in developing countries. *Paediatrics and International Child Health, 34*(4), 250–264.

Prentice, A. M., Ward, K. A., Goldberg, G. R., Jarjou, L. M., Moore, S. E., & Fulford, A. J. (2013b). Critical windows for nutritional interventions against stunting. Reply to JL Leroy et al. *American Journal of Clinical Nutrition.*

Prentice, A. M., Ward, K. A., Goldberg, G. R., Jarjou, L. M., Moore, S. E., & Fulford, A. J. (2013a). Critical windows for nutritional interventions against stunting. *American Journal of Clinical Nutrition, 97,* 911–918.

Ruel, M. T., & Alderman, H. (2013). The Maternal and Child Nutrition Study Group. 2013. Nutrition-sensitive interventions and programmes: How can they help to accelerate progress in improving maternal and child nutrition? *The Lancet, 382*(9891), 536–551. http://dx.doi.org/10.1016/S0140-67361360843-0.

Smith, L. C., & Haddad, L. (2015). Reducing child undernutrition: past drivers and priorities for the post-MDG era. *World Development, 68,* 180–204. https://doi.org/10.1016/j.worlddev.2014.11.014.

Stock, J. H., & Yogo, M. (2005). Testing for weak instruments in Linear IV regression. In J. H. Stock, & D. W. K. Andrews (Eds.), *Identification and inference for econometric models: Essays in Honor of Thomas J. Rothenberg*. New York: Cambridge University Press.

UNICEF. (1998). *The state of the world's children*. New York: United Nations Children's Fund.

Vollmer, S., Harttgen K., Subramanyam, M., Finlay, J., Klasen, S., & Subramanian, S. V. (2014). Association between economic growth and early childhood undernutrition: evidence from 121 Demographic and health Surveys from 36 low-income and middle-income countries. *The Lancet Global Health 2*, April. http://dx.doi.org/10.1016/S2214-109X1470025-7.

Webb, P., & Block, S. (2012). Support for agriculture during economic transformation: Impacts on poverty and undernutrition. *Proceedings of the National Academy Sciences, 109*(31), 12309–12314. https://doi.org/10.1073/pnas.0913334108.

World Bank. (2007). *World Development Report 2008: Agriculture for development*. World Bank, Washington D.C. https://doi.org/10.1596/978-0-8213-7235-7.

World Bank, World Development Indicators, The World Bank, Washington D.C. http://data.worldbank.org/data-catalog/world-development-indicators, Dataset, Accessed 1st February 2016.

World Health Organization, Geneva. http://apps.who.int/gho/data/node.main.522, Dataset. Accessed February 1, 2016.

7

Importance of Smallholder Farms as a Relevant Strategy to Increase Food Security

Peter Hazell

Abstract This chapter enumerates the importance of small farms for food security, pulling together available studies and quantitative evidence on the current status of small farms in terms of their number, share of total farms, share in farmed area, employment share, age, gender, poverty and food insecurity status, importance in marketed surpluses of food staples, income diversification, etc. To the extent possible, trends in these variables are also enumerated, focusing on key questions such as: are small farms getting smaller (not just the average size of all farms); are they are becoming less important in total food supplies, especially marketed surpluses (needed to feed the cities); how successfully do they use high-value agriculture and nonfarm income diversification to offset smaller farm sizes. To the extent possible, differences in patterns between regions and types of countries are identified. Finally, some scenarios for the future are developed.

1 Introduction

Small farm-led development has been the dominant agricultural development strategy since its remarkable success in driving Asia's Green Revolution. The paradigm is based on three major advantages claimed for small farms in poor, labour abundant countries.

- Small farms are more efficient than large farms, as evidenced by an impressive body of empirical studies showing an inverse relationship between farm size and land productivity across Asia and Africa (Binswanger-Mkhize and McCalla 2010; Eastwood et al. 2010; Larson et al. 2014). Amongst other things, this means small farms can produce more food than large farms in a limited land area.
- Small farms are the most populous farm size group, and they also farm large shares of the total agricultural area. As such, when scaled up, small farm development can have a big impact on agricultural growth and national food security.

P. Hazell (✉)
Santa Barbara, CA, USA
e-mail: p.hazell@cgiar.org

- Not only are small farms more efficient, but they also accounted for large shares of the rural poor. As such, small farm development can be a 'win-win' proposition for growth and poverty reduction.

There is much debate today about the continuing relevance of the small farm development paradigm. Important changes that may have eroded some of their perceived advantages include the following.

- With continued population growth, small farms have shrunk in size, making it harder to support a family from farming alone.
- Globalisation and market liberalisation policies have led to more consumer-driven food systems that are more dominated by large-scale players and place a greater emphasis on quality and safety attributes. Many small farms have difficulty connecting to such markets.
- The substantial state interventions to support small farms, such as fertiliser subsidies and guaranteed markets, were removed as part of market liberalisation policies, with the theory that the private sector would fill the gap. The private sector did expand, but often not at sufficient scale, and many small farms lost access to key inputs and markets.
- Many developing countries have reached middle-income status and are rapidly urbanising. One consequence is that rising labour costs are eroding some of the efficiency advantages of small farms and encouraging a shift towards more capital-intensive farming.

Despite these challenges and the pessimism of some authors (e.g. Maxwell et al. 2001; Collier 2009), small farms are not disappearing. Quite the opposite, they are multiplying, and in some countries are becoming even more dominant in the land distribution. This paper reviews recent changes in smallholder farming and explores some of the implications for food security.

2 Trends in the Number and Size of Small Farms

There are many possible definitions of small farms, but this chapter follows a common practice and defines smallholders as holdings of less than two hectares in size. A simple land-size definition enables the number of small farms to be enumerated using widely available agricultural census and household survey data, and facilitates comparisons across countries and over time. However, it can also be a misleading definition, because the economics of size depends on the quality of the available land, the prevailing agro-ecological conditions and the income-earning opportunities available to farmers. A 'viable' small farm might vary from just a couple of irrigated hectares of land in Asia to several hundred hectares of rainfed lands in parts of Latin America. But it can be much smaller if farming is combined with non-farm sources of income, as with many small, part-time farmers in Asia and Africa.

Some argue for a broader definition of smallholders, based on the concept of the family farm—farms that are operated by farm families using largely their own labour. Since just about every small farm is a family farm, this definition leads to higher estimates of the total number of 'small' farms. For example, in Latin America, there are about 5 million small farms less than 2 ha, but about 20 million family farms (Berdegué and Fuentealba 2014). A problem with using family farms as a definition of small farms is that they are not all small. In fact, many of the larger commercial farms found around the world are also family farms as defined above; they are just highly capitalised and mechanised ones. Attempts to draw a line between 'small' and 'non-small' family farms are sometimes made on the basis of assets (e.g. capital stock or machines), farm income or gross turnover. But this requires access to data that are not widely available, and definitions that are not easily comparable across countries and over time. The broader family farm approach is also less compelling for Asia and the more populous countries of sub-Saharan Africa (SSA), where the average farm size is well under 2 ha and there are relatively few large family farms.

Even with a land-based definition of size, obtaining reliable, recent and comparable cross-country data on trends in the number and size of small farms is still a challenge. The best data sources are national agricultural censuses since these adhere to established international guidelines and provide reliable information about the distribution of land as well as farm households (Lowder et al. 2016). The difficulty is that recent agricultural census data are not available for many countries, and most of the available data pertain to the late 1990s or early 2000s. More recent household surveys are available, such as the Living Standard Measurement Surveys (LSMSs), but because they are sampled to represent households and not agricultural area, they can seriously understate the importance of large farms, and hence distort estimated land shares. Also, most household surveys typically exclude farms that are not operated by farm families. Lowder et al. (2016) illustrate the problems by comparing census and household survey data from Guatemala, showing that while farms less than 3.5 ha accounted for 86% of all farms and 14% of the agricultural area in the 2003 agricultural census, a household survey conducted in 2006 found that the same size group accounted for 94% of all farms and 65% of the land area. Moreover, farms larger than 45.2 ha did not show up at all in the household survey but accounted for 57% of the land area in the 2003 agricultural census.

Based on agricultural census data from the 1990s and early 2000s, Lowder et al. (2016) estimate that there are about 570 million farms in the world, of which some 475 million (about 85%) are small (≤2 ha). About 90% of all farms are located in developing countries, and they are predominantly concentrated in Asia and Africa (60% of them are in China and India).

There has been a substantial increase in the number of small farms in recent decades. Since there does not seem to have been a comparable global estimate to Lowder et al. (2016) for an earlier period, it is necessary to fall back on census data for individual countries. Lipton (2009) reports changes between censuses for selected countries in Africa, Asia and Latin America and the Caribbean (LAC) (Table 1).

Except for Turkey, where the number of holdings and total area declined by about 20% between the two censuses, the number of holdings increased by between 12% (Uruguay) and 77% (Ethiopia) for all the other countries in Table 1; the unweighted average increase is 29%. The number of holdings of ≤2 ha also increased in all

Table 1 Changes in number and area of farms and smallholders, selected countries

Country	Year	Number holdings (million)	Area (million ha)	Number holdings ≤2 ha (million)	% holdings ≤2 ha	% area held by holdings ≤2 ha
Colombia	1988	1.45	36.03	0.51	35.6	1.7
	2001	2.02	50.71	0.83	41.1	2.1
Egypt	1990	2.91	3.30	2.61	89.9	48.9
	1999–00	3.72	3.75	3.53	95.0	57.5
Ethiopia	1988–92	6.09	4.87	5.62	92.3	62.1
	2001–02	10.76	11.05	9.37	87.1	60.4
India	1986	97.16	164.56	74.0	76.2	29.0
	1995–96	115.58	163.36	92.8	80.3	36.0
Nepal	1992	2.74	2.60	2.44	89.2	57.6
	2002	3.34	2.65	3.08	92.3	68.7
Pakistan	1990	5.07	19.15	2.40	47.4	11.3
	2000	6.62	20.41	3.81	57.6	15.5
Panama	1990	0.21	2.94	0.12	58.1	1.4
	2001	0.24	2.77	0.13	52.7	0.6
Thailand	1988	4.88	17.46	1.30	26.7	6.7
	1993	5.65	19.00	1.86	32.9	7.6
Turkey	1991	3.97	23.45	1.38	34.9	5.7
	2001	3.02	18.43	1.01	33.4	5.3
Uruguay	1990	0.05	15.80	na	na	na
	2000	0.06	16.42	na	na	na

Source Lipton (2009)

countries except Turkey. More recent data show that in India, the number of small farms ≤2 ha increased from 92.8 million in 1995–96 to 107.6 million in 2005–06, while their share in total farms increased to 83.3% (World Bank 2007).

The average size of farms has also shrunk in many countries, as shown in Table 2.

Although the overall trend is towards more and smaller farms, there are some regional and country variations in how the distribution of land is changing.

- Farms are finally starting to get larger on average in China (up from 0.57 to 0.60 ha over the period 2005–2010) (Huang et al. 2012), but the more general pattern across Southeast Asia is still towards smaller farms. In the Philippines, the average operational farm size fell from 3.6 ha in 1971 to 2.0 ha in 2002, and the share of small farms ≤1 ha increased from 13.6 to 40.1%. Indonesia and Thailand saw more modest declines of 15–20% in average farm sizes over similar periods, but little change in the share of small farms ≤1 ha in size (Otsuka 2013).
- In South Asia, the number of farms is still growing and average farm sizes are shrinking. In India, the average farm size approximately halved between 1971 and 2005–06, and in Bangladesh, the average operational farm size shrunk from 1.4 ha in 1976–77 to 0.3 ha in 2005 (Otsuka 2013).

Table 2 Census and survey-based estimates of trends in average farm size (hectares)

	1960s–1980s	2000s	Change (%)
Small farm developing countries			
Sub-Saharan Africa ($N = 14$)	2.9	1.9	−32
Land abundant SSA ($N = 9$)	3.0	2.9	−2.1
Land constrained SSA ($N = 5$)	2.3	1.2	−46.9
India	2.7	1.2	−57
Other S. Asia ($N = 4$)	2.5	1.1	−56
Indonesia	1.0	0.8	−20
China	0.7	0.6	−17
Other SE Asia ($N = 4$)	1.6	4.2	158
Middle East and N. Africa ($N = 9$)	7.6	5.4	−29
Commercialised agricultural economies			
South Africa	965.6	288.3	−70
Argentina	383.3	582.5	52
Brazil	70.7	68.2	−3.6
Other South America ($N = 7$)	97.3	89.7	−8
Western Europe ($N = 16$)	14.7	20.8	41
Canada	187.5	315.0	68
USA	157.6	169.3	7
Australia and New Zealand	1468.5	2070.3	41

Source Headey (2016)

- African countries vary widely in their population densities, and farms are about half the size in highly populated countries than in less populated countries (Table 2; Jayne et al. 2016). Farm sizes have also shrunk the most in the highly populated countries; from around 2.3 ha in the 1970s to 1.2 ha in the 2000s, compared to a decline from 3.0 to 2.9 ha in less densely populated countries (Table 2; Jayne et al. 2016). Based on repeat household surveys in eight African countries, Jirström et al. (2011) found that even over the six-year period 2002–2008, average farm size declined by 15% in Ghana, 35% in Mozambique, 13% in Tanzania and 10% in Zambia, but remained unchanged in Kenya and Malawi, and increased by 9% in Ethiopia and by 37% in Nigeria. The average change across the eight countries was a decline of 11% (from 2.4 to 2.2 ha per holding).
- In LAC, small family farms are typically larger than in Asia and Africa, and as defined by Berdegué and Fuentealba (2014), there are about 20 million of them (of which 5 million <2 ha). Their numbers are increasing in Central America but seem more stable across much of Latin America.

Small farms account for sizeable shares of the total farmed area in many Asian and African countries, but for much smaller shares in much of LAC (Table 1; Thapa and Gaiha 2014; Berdegué and Fuentealba 2014). Similarly, Lowder et al. (2016)

estimate that small farms ≤2 ha account for 75–80% of all holdings in South Asia, SSA, and East Asia and Pacific (excluding China), and 30–40% of the agricultural area. In LAC, they estimate that about 25% of the holdings are ≤2 ha. These are predominantly concentrated in Central America and the Andes, and while important in some countries, they farm only a tiny share of the total agricultural area in LAC (Lowder et al. 2016).

3 Trends in Small Farm Livelihood Strategies

Many small farms have reached the point where they are now too small to provide a full-time living for a household, and this has led farm households to diversify into high-value farming, wage employment, migration, and other non-farm activity. In China, non-farm income shares for farm households increased from 33.7% in 1985 to 63% in 2000 and 70.9% in 2010 (Huang et al. 2012). This is an extreme example, but non-farm income shares have reached 40% or more in many other Asian and SSA countries and are often much higher for the smallest farms (Haggblade et al. 2007). In India, diversification has helped prevent widening income gaps between rural and urban households (Binswanger-Mkhize 2012). On average, this diversification is higher across Asia than Africa, but there is considerable variation within each continent.

Diversification is enabled by rapid urbanisation, and the growth of small- and medium-sized towns. Already, 37% of the population in Africa, 48% in Asia, and 80% in LAC live in urban areas, and the United Nations projects that by 2050, the urban population shares are expected to reach 56% in Africa, 64% in Asia and 86% in LAC (UN 2014). Urbanisation is not just about megacities. In fact, close to half of the world's urban dwellers reside in relatively small settlements of less than 500,000 inhabitants, and these are often the fastest-growing urban areas. This has important implications for urban–rural linkages, enabling many households to combine farm and non-farm activities, even when living in urban areas. Often, they are living in areas that were previously defined as rural, and have become urban through census reclassification rather than any physical move.

4 Why the Slow Exodus of Small Farms?

Despite growth,—sometimes quite rapid growth—in national per capita incomes and urbanisation, we are not yet seeing the patterns of farm consolidation that occurred during the economic transformation of most of today's industrialised countries. Rather, the continuing growth in smaller and more diversified farms might best be described as a 'reverse transition' (Hazell and Rahman 2014).

There are many factors driving this reverse farm size transition.

- An important driver is rural population growth, especially growth in working-age adults. Other than China, few countries are generating enough jobs in urban-based manufacturing, which has historically been the primary absorber of rural–urban migrants. Instead, workers are moving into service sector jobs, many of which are based in medium- and smaller-sized towns, not just big cities, and this enables many rural households to diversify into non-farm employment from their farm base.
- Rapid urbanisation, globalisation and a shift towards more diversified diets are creating greater opportunities for some small farms to succeed by growing and marketing high value, labour-intensive livestock and horticultural products (Joshi et al. 2007; McCullough et al. 2008).
- The increasing use of subsidies and other agricultural support policies make small-scale farming more attractive than its real economic worth. China and many other Asian countries have already introduced farm support policies of various kinds, much as Japan, South Korea and Taiwan did during their economic transformation (OECD 2016).
- There are also many country-specific institutional and cultural constraints that act to keep people on the farms, including:
  - constraints on rural–urban migration, such as language, racial, and cultural barriers; legal restrictions on resettlement (e.g. China);
  - inheritance systems that lead to subdivision of farms among multiple heirs;
  - restrictions on land market transactions, such as caps on farm size (India) or indigenous land rights systems that limit opportunities for land consolidation (Africa);
  - religious and cultural constraints on women's employment opportunities other than farming;
  - inadequate social security systems, so that farms are kept as a retirement hedge.

Many of these drivers are very powerful and seem unlikely to diminish in the near future. Take rural population growth, for example. Rural populations are projected to nearly double by 2050 in Africa, so the pressure on land will keep growing (Fig. 1). Rural population growth is slowing in much of Asia and will approach tipping points by 2030, at which point the pressure on the land base will begin to reverse. This has already happened in Bangladesh and China and may be happening more widely in dynamic regions with good market access within countries (Masters et al. 2013).

5 Prognosis

Should we expect much change in these patterns over the next two to three decades? Much will depend on rates of national economic growth and the non-agricultural employment intensity of that growth. But rapid farm consolidation does not necessarily follow from economic growth, because of some of the constraints listed above.

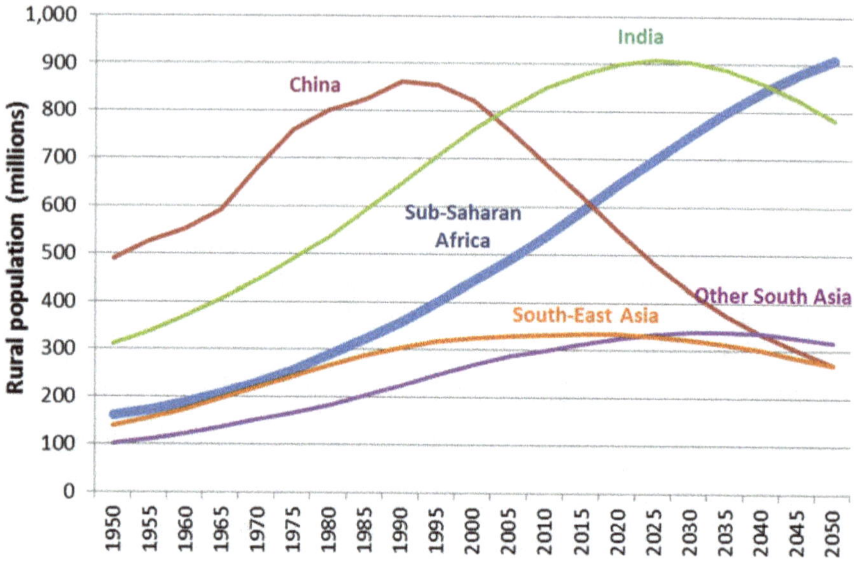

Fig. 1 Projected growth in rural populations. *Source* United Nations (2018)

Moreover, rapid urbanisation, and particularly the growth of small- and medium-sized towns, increases opportunities for farm households to diversify into non-farm sources of income from a farm base.

The earlier experiences of Japan, Taiwan and South Korea suggest that in Asia, the dominance of small farms could continue well into middle-income status (Otsuka 2013). In Japan, for example, the average farm size only started to increase quite recently despite the country's rapid economic take-off in the 1960s. The average farm size was still only 1.8 ha in 2005, and the percentage of farms ≤3 ha was still 90.5%.

Small farms not only dominate the total number of holdings in Asia and Africa, but as data from the 1990s and early 2000s show, they have typically retained large land shares, leaving little room for the emergence of many medium-sized and large farms. In some countries (e.g. Bangladesh, India and the Philippines), even the total agricultural land area is becoming more concentrated among small farms, and it is the large farms that are being squeezed out. On the other hand, there is evidence that some consolidation of operated, rather than owned, farmland is occurring in some countries, with small farms renting out some of their land to larger-scale operators (Otsuka 2013). In another development, Jayne et al. (2016) found that pockets of medium-sized farms are emerging in parts of Kenya, Malawi and Zambia, some of which are operated by urban-based investor farmers.

Another reason to think there may be greater land consolidation in the future is that the average age of farmers is increasing (currently about 60 years in Africa), and some land consolidation may eventually occur as part of an intergenerational transition. However, this may be offset by offspring who have left farming to work elsewhere but return to the farm when they retire, a not uncommon practice in many Asian and African countries. We simply do not know much about these possible demographic trends.

While things will eventually change, widespread land consolidation could be decades away. In the meantime, agricultural development is going to be largely all about small farms.

What we may see in the future is an increasing diversity of farm household livelihoods, with increasing gaps.

1. Gaps between commercially oriented small farms that are well linked to value chains and a much larger number of subsistence or non-farm-oriented farms. Christen and Anderson (2013) estimate that only about 35 million of the world's small farms (about 8%) participate in tight value chains, implying that the vast majority of small farms are being left behind.
2. Gaps between small farms in favourable areas with good market connectivity and those in poorly connected and often marginal areas (lagging regions). This has already happened in much of South Asia, where rural poverty is now concentrated in lagging regions (Ghani 2010).

This increasing diversity will be a challenge for future assistance programmes for small farms, and interventions will need to be more carefully targeted. Several small farm typologies have been proposed in the literature to help guide such strategies, which have been summarised into three classes of small farms (Dorward et al. 2009; Hazell and Rahman 2014):

- *Commercially oriented small farmers* already successfully linked to value chains, or who could link if given a little help. Many commercially oriented small farms are part-time farmers.
- *Small farms in transition,* who have favourable off-farm opportunities and are at various stages of exiting farming as a serious business.
- *Subsistence-oriented small farms* marginalised for a variety of reasons that are hard to change, such as being located in remote areas with limited agricultural potential, or representing elderly or infirm farmers. Many of the same factors also prevent them from accessing non-farm jobs and becoming transition farmers.

The relative importance of these three small farm groups varies widely from region to region. In a less favoured region of a slow-growing country—the worst of all possible worlds, and a situation all too prevalent in Africa—there are relatively few market-oriented farms, but many subsistence-oriented small farmers, including those who are trying to transition out of farming but cannot because of a shortage of off-farm opportunities. At the other extreme, in a dynamic region of a dynamic country—such as some of the coastal areas in China—many small farmers are producing lots of high-value products for the market, or are transitioning into better-paid opportunities in the industrial areas and in their local non-farm business economy. Relatively, few subsistence-oriented farmers remain, and these are often the elderly or the infirm. Many other regions, of course, fall somewhere between these two extremes.

With economic growth and urbanisation, significant numbers of commercially oriented small farms are likely to prosper through diversification into high-value agriculture. The most successful small farmers will tend to be located in areas with good agricultural potential and market access. Over time, some commercially oriented small farmers will become medium or large farms, while others will eventually become transition farmers or successfully exit farming to the non-farm economy. Transition farmers will either have, or will be able to develop, suitable skills and assets for undertaking non-farm activity, and they are likely to live in well-connected areas with access to off-farm opportunities. Their farming activities are likely to be oriented towards their own consumption rather than the market. Subsistence-oriented farmers are more likely to persist in less favoured and tribal areas and to grow traditional food staples (both crop and livestock) for their own consumption.

Hazell and Rahman (2014) discuss the kinds of interventions that may be relevant for each of the three groups of small farms. Commercially oriented small farms need support as farm businesses. They need access to improved technologies and natural resource management (NRM) practices, modern inputs, financial services and markets, and secured access to land and water. Much of this assistance will need to be geared towards high-value production and provided on a business basis. Many smallholders will also require help acquiring the necessary knowledge and skills to become successful business entrepreneurs in today's value chains, especially women and other disempowered groups. Managing market and climate risk are challenges for many small farms; in addition to insurance and access to safety nets, these farms need to develop resilient farming systems.

Transition farmers need help for developing appropriate skills and assets to succeed in the non-farm economy, including, in many cases, assistance in developing small businesses. This can be especially important for women and other disempowered groups who have little experience working off-farm. The transition to the non-farm economy might also be facilitated by securing land rights and developing efficient land markets, so that transition farmers can more easily dispose of their farms. Since many transition farmers seem likely to continue to remain as part-time farmers, they can also benefit from improved technologies and NRM practices that improve their farm productivity.

Subsistence farmers are predominantly poor and will mostly need some form of social protection, often in the form of safety nets, food subsidies or cash transfers. Interventions that help to improve the productivity of their farms (e.g. better technologies and NRM practices) can make important contributions to their own food security, perhaps provide some cash income, and in many cases may prove more cost effective than some forms of social protection. Subsistence farmers have limited ability to pay for modern inputs or credit; however, so intermediate technologies that require few purchased inputs may be needed, or inputs will need to be heavily subsidised. Subsistence farmers are typically the most exposed and vulnerable to climate risks, and in addition to safety nets, they need help developing resilient farming systems.

6 Some Implications for Food Security and Nutrition

During the Green Revolution in Asia, small farms produced important amounts of the food staples that fed the cities as well as the countryside. Today, small farms are much smaller in size, and many are net buyers rather than net sellers of food staples. They are still able to provide for the food security of huge numbers of rural people, but it seems likely they contribute relatively little towards feeding urban populations.

Empirical evidence on these relationships is scarce. Available data on the share of marketed surpluses supplied by different farm size groups (e.g. Jayne et al. 2016) do not necessarily show this decline, because they do not differentiate between sales consumed in rural areas and sales consumed in urban areas. However, a recent study by Herrero et al. (2017) using spatial referenced data for 2005 provides some relevant insights. They show that, on average, small farms ≤2 ha produce about 55% of total cereals in China, 30% in Africa and Asia (excluding China), 10% in West Asia and North Africa, 15% in Central America, and negligible shares in South America (Table 3). Middle-sized farms of 2–20 ha are more important producers everywhere, except China, and large farms >20 ha dominate in Central and South America.

The importance of these results is that the small farm share in total cereal production in each region is considerably less than the corresponding rural population share (last column of Table 3), implying that in aggregate, small farms are no longer able to meet all the food staple needs of rural populations, let alone feed growing urban populations. It can be inferred that on balance, urban populations are now being fed mainly by medium-sized and large farms and from imports. In China, the small farm share in cereals production is still larger than the rural population share, so small farms may still be important suppliers to the urban market. Similar regional findings by Herrero et al. (2017) hold for the production of other food staples such as roots and tubers.

Table 3 Share of cereals produced in each region by farm size group, 2005

Region	Farm size group[a]			Estimated rural population share in 2015 (%)[b]
	≤2 ha	2–20 ha	>20 ha	
Sub-Saharan Africa	30	50	20	61
South Asia	30	55	15	66
Southeast Asia	25	55	20	53
China	55	15	30	44
West Asia and North Africa	10	45	45	West Asia 29 North Africa 49
Central America	15	30	55	26
South America	0	5	95	16

Source a Herrero et al. (2017)
b United Nations (2018)

Urban population shares are projected to grow strongly across the developing world, and feeding these populations will require even more rapid growth in marketed food surpluses. It follows that for many countries, a national security agenda for food staples needs three pillars. One pillar is to provide support to the many smallholders who farm largely to meet their own subsistence needs. A second pillar is to support those farms that can produce marketed surpluses for the cities, and which are increasingly likely to be medium- and larger-scale farms. The third pillar is an appropriate international trade policy to enable imported food staples to fill remaining gaps.

Many small farms, particularly those located in well-connected areas, do have a comparative advantage in growing labour-intensive, high-value products like fruits, vegetables and milk, and this means they can contribute to more diversified and nutritionally rich diets in both rural and urban areas.

7 Conclusions

Overall, there are more small farms than ever today, and they are getting smaller, less important for supplying urban areas with food staples, and more dependent on non-farm sources of income for their livelihoods. However, there is considerable country and regional variation around this broad narrative. Small farms are not getting smaller everywhere, and in some countries, land is beginning to be consolidated into larger-sized holdings, even if only for operational purposes. Some small farms are also successfully marketing high-value perishable products like fruits, vegetables and milk, though relatively few are successfully linking into modern value chains. Non-farm income diversification is proving a successful livelihood strategy for small farms in fast growing countries and urbanising regions where more opportunities abound, but for many others, it is little more than a coping strategy that prevents or slows the descent into deeper poverty.

Several conclusions can be drawn from this analysis. First, small farm assistance programmes need to be cognisant of the growing diversity of small farm situations today, and to build strategies appropriate to each. These may need to distinguish between subsistence-oriented and market-oriented small farms, and small farms that are at various stages of transition out of farming through non-farm income diversification. It may also be necessary to differentiate between small farms that live in dynamic versus lagging regions, because of the different opportunities and constraints they face. This targeting requires the development and use of small farm classification schemes or typologies, and their operationalisation through appropriate mapping techniques (Hazell and Rahman 2014).

Second, national food security strategies need to be structured around three pillars. One pillar is to provide support to the many smallholders who farm largely to meet their own subsistence needs. The second pillar is to support investment in those farms, mostly medium- and large-scale farms, that can produce the marketed surpluses needed to feed the cities. The third pillar is an appropriate international trade policy to enable imported food staples to fill remaining gaps. The first of these pillars requires active public sector and NGO engagement with small farms, while the second pillar requires a proactive business agenda led by private sector initiatives.

Third, small farms have an important role to play in providing more diverse and nutritionally rich diets, both in producing a diverse array of food for their own and local market needs, but also help to provide urban areas with higher value and labour-intensive foods.

References

Berdegué, J. A., & Fuentealba, R. (2014). The state of smallholders in agriculture in Latin America. In P. Hazell & A. Rahman (Eds.), *New directions for smallholder agriculture*. Oxford: Oxford University Press.

Binswanger-Mkhize, H. (2012). India 1960–2010: Structural change, the rural non-farm sector, and the prospects for agriculture. *Center on Food Security and the Environment, Stanford Symposium Series on Global Food Policy and Food Security in the 21st Century*, Stanford, CA: Stanford University.

Binswanger-Mkhize, H., & McCalla, A. F. (2010). The changing context and prospects for agricultural and rural development in Africa. In P. Pingali & R. Evenson (Eds.), *Handbook of agricultural economics* (Vol. 4, pp. 3571–3712). Amsterdam: Elsevier.

Christen R., & Anderson, J. (2013). Segmentation of smallholder households: Meeting the range of financial needs in agricultural families, Focus Note No. 85, CGAP, April.

Collier, P. (2009). Africa's organic peasantry: Beyond romanticism. *Harvard International Review*, Summer 62–65.

Dorward, A., Anderson, S., Nava, Y., Pattison, J., Paz, R., Rushton, J., et al. (2009). Hanging in, stepping up and stepping out: livelihood aspirations and strategies of the poor. *Development in Practice, 19*(2), 240–247.

Eastwood, R., Lipton, M., & Newell, A. (2010). Farm size. In P. Pingali & E. Robert (Eds.), *Handbook of agricultural economics, Volume 4,* (pp. 3323–3397). Elsevier: Amsterdam.

Ghani, E. (Ed.). (2010). *The poor half billion in South Asia: What is holding back lagging regions?* New Delhi: Oxford University Press.

Haggblade, S., Hazell, P., & Reardon, T. (Eds.). (2007). *Transforming the rural nonfarm economy*. Baltimore: Johns Hopkins University Press.

Hazell, P., & Rahman, A. (2014). Concluding chapter: The policy agenda. In P. Hazell & A. Rahman (Eds.), *New directions for smallholder agriculture* (pp. 527–558). Oxford: Oxford University Press.

Headey, D. (2016). The evolution of global farming land: Facts and interpretations. *Agricultural Economics, 47*(S1), 185–196.

Herrero, M., Thornton, P. K., Power, B., Bogard, J., Remans, R., Fritz, S., Gerber, J, Nelson, G., See, L., Waha, K., Watson, R., West, P., Samberg, L., van de Steeg, J., Stephenson, E., van Wijk, M., & Havlik, P. (2017). *Farming and the geography of nutrient production for human use: A transdisciplinary analysis. Lancet Planet Health, 1*(1): e33–42. http://thelancet.com/pdfs/journals/lanplh/PIIS2542-5196(17)30007-4.pdf.

Huang, J., Wang, X., & Qui, H. (2012). *Small-scale farmers in China in the face of modernisation and globalisation*. London/The Hague: International Institute for Environment and Development (IIED)/Hivos.

Jayne, T. S., Chamberlin, J., Traub, L., Sitko, N., Muyanga, M., Yeboah, F. K., Anseeuw, W., Chapoto, A., Wineman, A., Nkonde, C., Kachule, R. (2016). Africa's changing farm size distribution patterns: the rise of medium-scale farms. *Agricultural Economics, 47,* 197–214.

Jirström, M., Andersson, A., & Djurfeldt, G. (2011). Smallholders caught in poverty—Flickering signs of agricultural dynamism. In G. Djurfeldt, E. Aryeetey, & A. Isinika (Eds.), *African smallholders: Food crops, markets and policy* (pp. 74–106). Wallingford, Oxfordshire: CABI.

Joshi, P. K., Gulati, A., & Cummings, R., Jr. (Eds.). (2007). *Agricultural diversification and smallholders in South Asia*. New Delhi, India: Academic Foundation.

Larson, D., Otsuka, K., Matsumoto, T., & Kilic, T. (2014). Should African rural development strategies depend on smallholder farms? An exploration of the inverse-productivity hypothesis. *Agricultural Economics, 45*(3), 335–367.

Lipton, M. (2009). *Land reform in developing countries*. London and New York: Routledge.

Lowder, S. K., Skoet, J., & Raney, T. (2016). The number, size, and distribution of farms, smallholder farms, and family farms worldwide. *World Development, 87*, 16–29.

Masters, W. A., Djurfeldt, A., De Haan, C., Hazell, P., Jayne, T., Jirström, M., Reardon, T. (2013). Urbanization and farm size in Asia and Africa: Implications for food security and agricultural research. *Global Food Security, 2*(3), 156–165.

Maxwell, S., Urey, I., & Ashley, C. (2001). *Emerging issues in rural development: An issues paper*. London: Overseas Development Institute.

McCullough, E. B., Pingali, P. L., & Stamoulis, K. G. (Eds.). (2008). *The transformation of agri-food systems; globalization, supply chains and smallholder farms*. London and Sterling, Virginia: FAO and Earthscan.

OECD. (2016). Producer and consumer support estimates. Online database Accessed at www.oecd.org.

Otsuka, K. (2013). Food insecurity, income inequality, and the changing comparative advantage in world agriculture. *Agricultural Economics, 44*(s1), 7–18.

Thapa, G., & Gaiha, R. (2014). Smallholder farming in Asia and the Pacific: Challenges and opportunities. In P. Hazell & A. Rahman (Eds.), *New directions for smallholder agriculture* (pp. 527–558). Oxford: Oxford University Press.

United Nations. (2018). *World urbanization prospects; the 2018 revision*. United Nations, New York: Department of Economic and Social Affairs.

World Bank. (2007). *World development report 2008: Agriculture for development*. Washington, DC: The World Bank.

8

Transforming Smallholder Agriculture to Achieve the SDGs

Mathew Abraham and Prabhu Pingali

Abstract There is overwhelming historical evidence from the developed world and from the newly emerging economies of the developing world that indicates that agricultural growth has been the primary engine of overall economic growth. The transformation of economies around the world, from predominantly agricultural to industrial, was kick-started by rapid agricultural productivity growth. Does the growth in agricultural productivity have to necessarily come from the small farm sector? Rapid improvement in small farm productivity is one of the primary mechanisms by which dramatic rural poverty reductions can be achieved as shown by the Green Revolution experience in Asia and more recently in sub-Saharan Africa. Economic growth policies that are inclusive of smallholder farmers directly contribute to the SDG 2 that is focused on ending hunger, achieving food security and promoting sustainable agriculture. Past efforts at small farm productivity improvement were focused on staple grains, looking ahead one needs to take a food systems perspective and encourage diversification into nutrition-rich legumes, pulses, horticulture crops and livestock. Investment in rural market infrastructure allows smallholders to commercialise and enhance the supply of perishable products. Linking small farms to urban food value chains is also a promising new avenue for rural poverty reduction.

1 Introduction

In 2015, the United Nations approved the 2030 Agenda for Sustainable Development, setting in motion the Sustainable Development Goals (SDGs). The 17 goals of the SDGs have 169 targets, designed to take a holistic approach to addressing the social, economic and environmental aspects of sustainable development. Although

M. Abraham (✉) · P. Pingali
Tata-Cornell Institute for Agriculture and Nutrition, The Charles H. Dyson School of Applied Economics and Management, Cornell University, Ithaca, New York, USA
e-mail: ma947@cornell.edu

P. Pingali
e-mail: plp39@cornell.edu

Goal 2 of the SDGs—which aims to end hunger and malnutrition and double agricultural productivity and incomes of small-scale farmers—is directly linked to small farm production, eight other goals related to ending poverty, gender discrimination, inequality and environmental degradation, tackling climate change, and promoting and ensuring healthy lives have small farm development and growth central to their success. A majority of the world's agricultural production takes place on small farms, and currently 90% of the 570 million farms globally are small (less than 2 ha in size) and cultivated by 1.5 billion of the world's poor (Rapsomanikis 2015). In Asia and sub-Saharan Africa (SSA), where the problems of hunger and poverty are most severe, 80% of food supply comes from smallholders[1]. Therefore, assuring the viability of small farms is crucial to meeting the SDGs.

Small farms face numerous challenges in production, especially in terms of access to essential factors of production, such as credit, inputs (seeds, fertilisers, pesticides), information and production technologies, in addition to poor access to output markets (Pingali 2012; Poulton et al. 2010). Small farms are heterogeneous economic units of agricultural production. Their characteristics and challenges vary according to geography, the influence of historical institutions and the political and socio-economic conditions in which they are situated. Therefore, addressing the concerns of small farm productivity and designing potential solutions to address them will vary within and across countries. Improving agricultural productivity and household-level incomes are central to reducing the poverty and nutritional challenges we face globally.

The aim of this chapter is to identify the various challenges in small farm economies at various stages of structural transformation, and the major interventions that are needed to improve their productivity in the context of meeting the SDG of ending poverty and ensuring prosperity for all. In the first part of the chapter, we identify the various goals of the SDGs that explicitly depend on small farm growth for their achievement, bringing to light the importance and urgency of interventions in small farm production systems. Productivity in small farms is influenced by geography, sociopolitical conditions and policy, and the farms vary in economies at different levels of structural transformation. In the second part of this chapter, we look at the major characteristics of low-productivity agricultural systems (much of SSA), modernising agricultural systems (South Asia and Latin America) and commercialised agricultural systems (East Asian economies; mainly South Korea, Taiwan and Japan). Here, we try to ascertain how challenges to the development of smallholder agriculture differ in each region, to make a case for context-specific interventions to achieve the SDGs.

In the last part, using a transaction cost framework, we try to understand the major challenges small farms face in different production systems, and we explain how these challenges may hinder farm viability. We also discuss how transaction costs, at both farm level and market level, may influence the incentives and capacity of different production systems to innovate, produce and sell, which forms the basis

[1]http://www.fao.org/fileadmin/templates/nr/sustainability_pathways/docs/Factsheet_SMALLHOLDERS.pdf.

for the development and growth of small farms. Here, we try to assess the various policy and institutional interventions that have the potential to mitigate transaction costs, at both the production and marketing stages, in different regions. We argue that in different agricultural systems, different sets of interventions are necessary to enable small farm growth.

2 Achieving the SDGs and the Centrality of Small Farms

The World Food Summit (WFS) goals and the Millennium Development Goals (MDGs) framework were the first systematic global attempts to monitor progress towards hunger reduction through internationally agreed benchmarks. The WFS took place in Rome in 1996, with representatives from 182 nations pledging, '…to eradicate hunger in all countries, with an immediate view to reducing the [*absolute*] number of undernourished people to half their present level no later than 2015'. The number of undernourished is those who fall below the minimum level of dietary consumption for a given country and year. Five years later in 2001, the UN as part of its MDG framework established a second benchmark, by which representatives of 189 nations pledged to fight extreme poverty in its many dimensions and 'to halve, between 1990 and 2015, the *proportion* of people whose income is less than one dollar a day … [and] … the proportion of people who suffer from hunger'. This became the first of the eight MDGs. Its central aim was to halve the prevalence of undernourishment, or the proportion of people below the minimum level of dietary consumption, between 1990 and 2015 (MDG 1c).

Although the MDG target of halving the prevalence of hunger was met, the WFS goal of halving *absolute* numbers of hungry was not accomplished (Pingali 2016). That said, the WFS goals and the MDGs did play a crucial role in shaping global thinking and action around poverty and hunger, paving the way for a bolder set of development goals. In 2015, UN member states approved the 2030 Agenda for Sustainable Development—to be achieved through 17 SDGs—in order to 'build on the work of the MDGs and complete what they did not achieve'.[2] Given that the SDGs were designed to take a holistic approach to addressing the social, economic and environmental aspects of sustainable development, that they were developed through an extensive consultative process, and that they are riding on the momentum of the MDG experience, the SDGs present a good opportunity for the world to continue the progress made in the MDG era. This is a bold vision. We believe the transformation of smallholder agriculture is critical to the task.

To summarise, Table 1 lists the 17 SDG goals and targets the global community has pledged to address by the year 2030. Out of the 17 goals, 9 goals (*italics*) directly pertain to the agricultural sector and have relevance to small farm growth and development.

[2]United Nations (2015).

Table 1 The 17 sustainable development goals

	Goal	Target
1	*No poverty*	*End poverty in all its forms everywhere*
2	*Zero hunger*	*End hunger, achieve food security and improved nutrition and promote sustainable agriculture*
3	*Good health and well-being*	*Ensure healthy lives and promote well-being for all at all ages*
4	Quality education	Ensure inclusive and equitable quality education and promote lifelong learning opportunities for all
5	*Gender equality*	*Achieve gender equality and empower all women and girls*
6	Clean water and sanitation	Ensure availability and sustainable management of water and sanitation for all
7	Affordable and clean energy	Ensure access to affordable, reliable, sustainable and modern energy for all
8	*Decent work and economic growth*	*Promote sustained, inclusive and sustainable economic growth, full and productive employment and decent work for all*
9	Industry, innovation and infrastructure	Build resilient infrastructure, promote inclusive and sustainable industrialisation and foster innovation
10	*Reduced inequalities*	*Reduce inequality within and among countries*
11	Sustainable cities and communities	Make cities and human settlements inclusive, safe, resilient and sustainable
12	*Responsible consumption and production*	*Ensure sustainable consumption and production patterns*
13	*Climate action*	*Take urgent action to combat climate change and its impacts*
14	Life below water	Conserve and sustainably use the oceans, seas and marine resources for sustainable development
15	*Life on land*	*Protect, restore and promote sustainable use of terrestrial ecosystems, sustainably manage forests, combat desertification, and halt and reverse land degradation and halt biodiversity loss*
16	Peace, justice and strong institutions	Promote peaceful and inclusive societies for sustainable development, provide access to justice for all and build effective, accountable and inclusive institutions at all levels
17	Partnership for the goals	Strengthen the means of implementation and revitalise the global partnership for sustainable development

Source https://sustainabledevelopment.un.org/sdgs

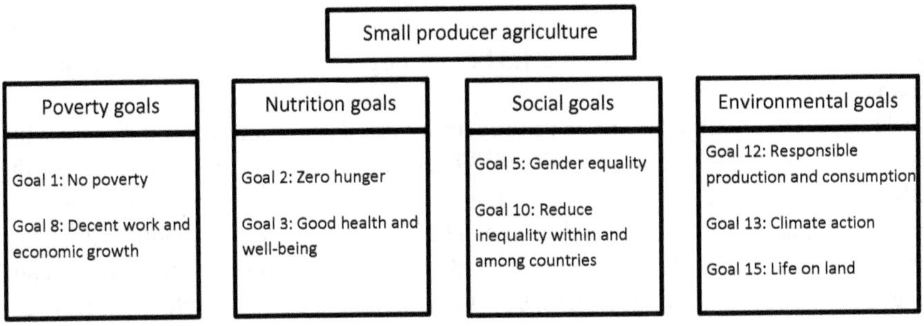

Fig. 1 Various goals for small producer agriculture development and growth

With over 1.5 billion people living in small producer households globally, their development is crucial for income growth, poverty reduction, food security, gender empowerment and environmental sustainability (Byerlee et al. 2009; Pingali 2010). Therefore, the growth and development of small producer agriculture are central to meeting the SDGs. In developing countries, multiple stressors (climatic as well as political), economic and social conditions influence food security (Leichenko and O'Brien 2002). In order to contextualise the SDGs and small farm development, we categorise SDGs into poverty, nutritional, social and environmental goals (Fig. 1). In the following subsections, we look at each of these goals in the context of small farm development and growth, to assess the major challenges in achieving them.

2.1 *Poverty Goals*

The SDG to end poverty in all its forms everywhere is especially targeted at over 836 million people who live on less than USD 1.25 a day. With a majority of the poor engaged in the agricultural sector, its growth and development are central to achieving this goal. Access to natural resources, property rights, basic services (R&D, finance) and risk reduction (price and climatic) become crucial for improving agricultural production. The role of agriculture development in poverty reduction is well established in economics literature. There is overwhelming evidence that, with very few exceptions, sustained reduction in poverty cannot be achieved without productivity increases in the agricultural sector (Timmer and Akkus 2008). Time series data used in various studies have shown the marginal effects of agricultural GDP growth on poverty reduction to be significant. Thirtle et al. (2003) estimated that with a 1% increase in crop productivity in Asia, poverty reduced by 0.48%. In the context of India, Fan et al. (2000) show a decrease of 0.24% in poverty with 1% growth in agricultural productivity. The experience in China, where there is more equitable land distribution, shows that growth in agricultural GDP led to four times higher reduction in poverty than in the non-agricultural sector (Ravallion and Chen 2007). In low-income countries, Christiaensen et al. (2006) find a 2.3 times larger increase in poverty reduction with agricultural growth. The same study shows a 4.25 times larger increase in SSA and a 1.34 times larger increase in the middle-income countries of North Africa.

Thus, growth and development of the agricultural sector are central to achieving the poverty goals (SDG 1 and SDG 8). These goals are also interlinked with the other group of goals identified in Fig. 1, as improved income is crucial to improving access to nutritious food, to end hunger and to reduce inequality both within and between countries. Reducing social inequality through empowerment of women and marginalised groups expands access to resources and services, which in turn can improve farm-level productivity. The urgency of climate action and conservation is also significant and inextricably linked to agricultural production. Along with increasing and sustaining growth, ensuring responsible production and consumption is important for reducing externalities such as emissions, soil degradation, water contamination and climate change, which ultimately put agricultural production at risk.

2.2 *Nutrition Goals*

The goal to end hunger, achieve food security and improve nutritional status is urgent. According to the FAO, 795 million people globally are undernourished; a majority of them live in Asia and about 281 million reside in SSA (FAO, IFAD and WFP 2015). Micronutrient deficiencies were a major issue that was under-addressed in the MDG and WFS goals of halving the prevalence and instances of the hungry (Pingali et al. 2016). The SDGs, however, are explicit in their aim to improve nutrition and to end all forms of malnutrition, focusing especially on wasting and stunting and also on the needs of adolescent girls, pregnant and lactating women and older people[3]. With over 3.1 million child deaths each year due to poor nutrition, and 66 million primary school children hungry (23 million in Africa alone), this is a major challenge. Although the number of stunted children under 5 declined from 225 million to 159 million globally, Africa and Oceania saw a 23 and 67% respective increase in the number of stunted children (ibid.). In terms of women's health, maternal mortality is also a serious concern globally. World Health Organization (WHO) data show that maternal mortality is 14 times higher in developing countries, and much of it is nutrition-based and preventable. In addition, about 42% of pregnant women in developing countries are anaemic (Kraemer and Zimmermann 2007), a condition which contributes to 20% of all maternal deaths.

Smallholder agricultural production is closely linked with nutrition and food security in three ways. Firstly, it makes food available through production; secondly, it reduces the real cost of food, making it more affordable; and thirdly, it improves incomes of farming households, enabling them to access nutritious foods (Ivanic

[3]Sustainable Development Goals Knowledge Platform—https://sustainabledevelopment.un.org/sdg2.

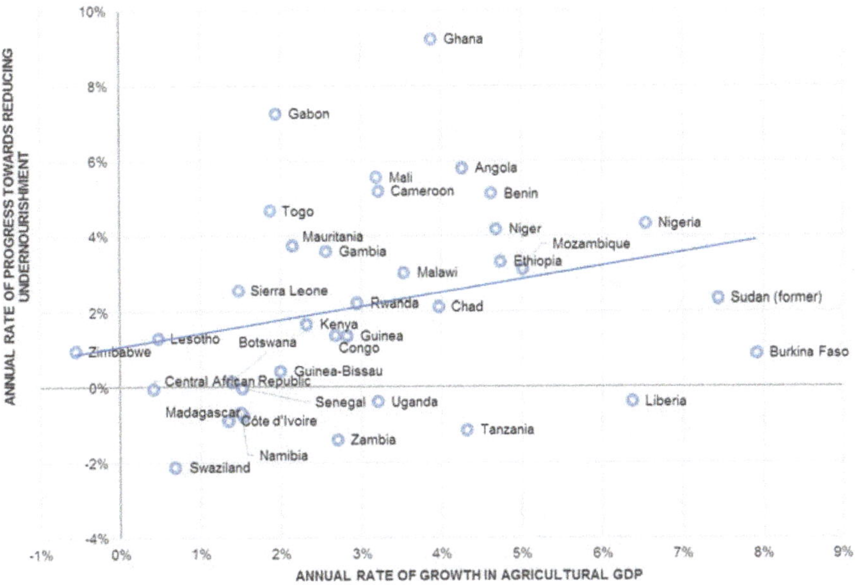

Fig. 2 Agricultural growth and reduction in hunger prevalence in SSA. *Source* Authors' analysis using FAO data for 2015

and Martin 2008; Pingali et al. 2015; Swinnen and Squicciarini 2012). Sufficient evidence exists to validate the relationship between agricultural growth and nutritional outcomes. Countries that proactively support pro-agricultural growth policies witnessed lower incidence of child stunting compared to countries that did not (Webb and Block 2012). FAO data substantiate this claim. For example, our analysis using FAO data shows that increases in agricultural growth correlate with decreases in hunger, stunting and child mortality in SSA (see Figs. 2, 3 and 4, respectively.) In other regions too, when there is support for sustaining agricultural development through policies targeting small producers, the resultant greater affordability of food has led to a decline in stunting and wasting. Meanwhile, countries with low agricultural productivity have consistently performed poorly on all three indicators (ibid.).

2.3 *Social Goals*

Achieving the social goal of reduced social inequality—especially gender inequality—depends on improved access to economic resources including land, natural resources, financial services and technology, for women and marginalised groups. Emancipation of these groups will be important to improve agricultural productivity, reduce regional inequalities and achieve sustained income growth. Women comprise 43% of the total agricultural labour force across the globe (FAO 2014), although

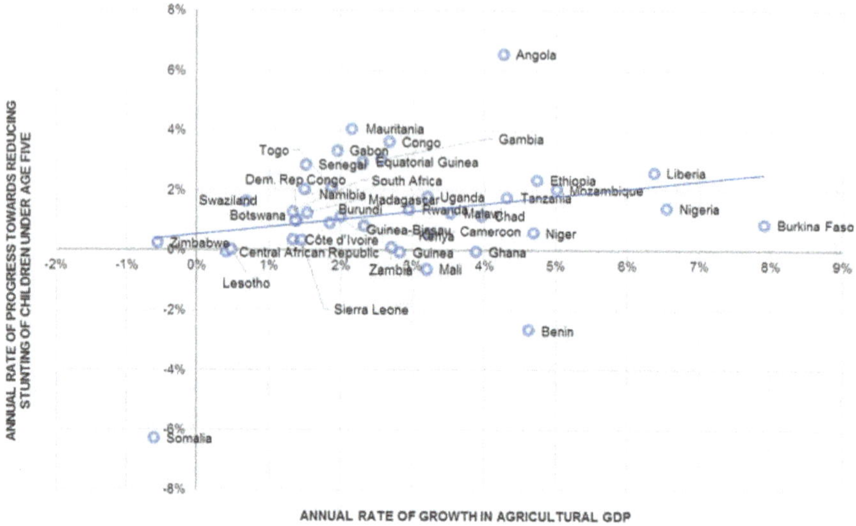

Fig. 3 Agricultural growth and reduction in prevalence of child stunting in SSA. *Source* Authors' analysis using FAO data for 2015

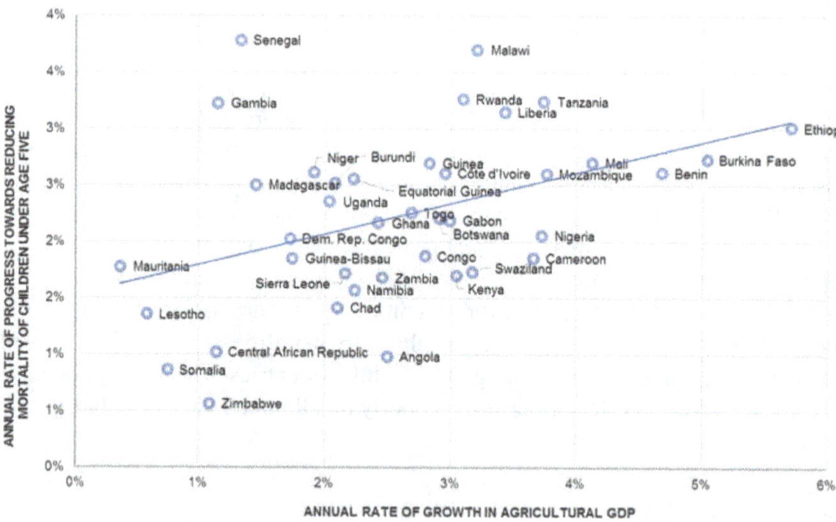

Fig. 4 Agricultural growth and reduction in prevalence of child mortality in SSA. *Source* Authors' analysis using FAO data for 2015

there are variations in this composition across the developing world. In SSA and in Southeast and East Asia, the percentage of women in agriculture is 50%, while in South Asia it is 35% and in Latin America, a little over 20% (FAO 2011). Women also make up over 66% of the 600 million small livestock managers (Distefano 2013).

It has been well established that there are high gender gaps, to the disadvantage of women, in access to and control of resources, especially land (Goldstein and Udry 2005; Quisumbing and Pandolfelli 2010), labour (Fontana 2009; Tzannatos 1999), credit (Sheahan and Barrett 2014), infrastructure, information and technology (Carr and Hartl 2010; Jost et al. 2016; Perez et al. 2015). This is largely due to institutional and norm-based constraints women face in society (Croppenstedt et al. 2013). The FAO (2011) reports that the underperformance of the agricultural sector is in part due to this differential access to resources for women, who represent a crucial aspect of production. Croppenstedt et al. (2013) conclude that fewer women (compared to men) are involved in the more profitable aspect of agriculture, i.e. commercial production.

2.4 Environmental Goals

The environmental goals, including climate action (SDG 13), responsible production and consumption (SDG 12), and the management and preservation of natural resources and biodiversity (SDG 15) are integral to small farm development. Temperature rises, and the unpredictability of floods, droughts and other extreme weather events resulting from climate change, influence the costs and conditions in which agricultural production takes place. At the same time, managing the environmental externalities of agricultural production, such as greenhouse gas (GHG) and non-GHG emissions, groundwater depletion and soil degradation, are also important concerns to increase food production for a growing population. Sustainable production and consumption therefore become an integral part of mitigation and adaptation strategies in the fight against climate change and wastage (especially for food and natural resources).

Changes in temperature increase the risks of pest attacks and disease outbreaks (O'Brien et al. 2004). This increases the cost of cultivation, due to the need for pest and disease management, and also escalates the risks of crop failure. Morton (2007) states that even a slight increase in temperature affects the conditions under which the major staples such as wheat, rice and maize are grown. Livestock production will also be impacted by climate change, posing significant and diverse challenges for food security. Quality and quantity of feed crop and forage, water availability, animal and milk production, livestock diseases and biodiversity are all important factors that will affect animal husbandry (Rojas-Downing et al. 2017). Temperature rise and humidity have an additional impact on food safety as they increase the risk of mycotoxin contamination in cereals and pulses (Paterson and Lima 2010), and of contamination of drinking water (Paerl and Huisman 2009), which in turn impacts nutrition outcomes (SDG 2 and SDG 3).

The externalities of agricultural production on the environment are also an important issue to consider in terms of climate change and of safeguarding the ecosystem. Deforestation, desertification, biodiversity loss and land degradation result from agricultural and infrastructure development. The agricultural sector accounts for 10–12%

of global anthropogenic GHG emissions, and it is also the main source of non-carbon dioxide GHGs, such as methane (CH_4—50%) and nitrous oxide (N_2O—60%) (Smith et al. 2007; Tubiello et al. 2013). The main sources of CH_4 emissions are enteric fermentation in livestock, anaerobic fermentation from inundated paddy, and livestock manure management. The major sources of N_2O are animal manure, synthetic fertilisers and crop residues. Myers et al. (2014) research using experimental data shows that high levels of CO_2 in wheat and rice cultivation decreases the grains' micronutrients, such as zinc (by 9.3%), iron (5.1%) and protein (7.8% in rice and 6.3% in wheat). Their research also shows that there was a small decrease in protein in field peas. This points to another potential impact climate change may have on nutrition (SDG 2).

Intervention and adaptation need to go together to mitigate the effects of climate change on smallholder agriculture and reduce the externalities from agricultural production. Extension services to improve agronomic practices and access to technology and infrastructure are important for smallholders to know how to adapt. Ensuring reduced wastage of food and food products, at the farm level and along the value chain, is also important to reduce production pressure and increase accessibility of food.

Overall, the growth and development of small producer agriculture systems are vital for meeting the poverty, nutrition, social and environmental goals. However, small producers are faced with significant challenges and constraints, characterised by poor access to production factors and agricultural commodity markets. High transaction costs in accessing goods and services hinder income growth and access to food, while increasing social pressures towards exploitation and drudgery to reduce labour costs. Environmental pressures, leading to land degradation and increased emissions, also result from low access to technology and poor agronomic practices that exert stresses on land to maximise returns. Understanding these challenges and constraints is crucial in enabling small farm development and growth. In the next section, we look at the specific characteristics of small farms and the major transaction costs they incur that limit their viability.

3 Economics of Small Farms and Stages of Structural Transformation

A majority of the world's agricultural production takes place on small and marginal farms and despite recurring predictions that small farms will soon disappear, they have persisted and in many cases have increased in number (Hazell et al. 2010). There are dramatic variations across the globe in landholding sizes and growth trends. SSA, South Asia, Southeast and East Asia largely comprise small farms with less than 2 ha of land, while in Europe and North America, landholdings are larger, averaging over 10 ha (Eastwood et al. 2010). Data from the 1970s onward show farms in North America, Europe and Oceania showing consolidating trends, while farms in Asia and

Table 2 Trends in landholding sizes in selected countries in selected years

Country	Period I		Period II		
	Year	Farm size (ha)	Year	Farm size (ha)	Percentage change
Canada	1971	187.6	2011	315	68.44
USA	1969	157.6	2012	175.6	5.09
Australia	1970	1920.3	2001	3232.1	68.31
France	1970	22.07	2010	54.6	140.6
Spain	1972	17.83	2010	24	23.38
Japan	1970	1	2002	1.57	57.00
Korea Rep.	1970	0.88	2002	1.4	59.09
Peru	1971–72	16.92	1994	20.1	18.79
Brazil	1970	59.4	1996	72.8	22.56
Guatemala			2006	1.06	
India	1971–72	2.3	2011	1.10	−52.17
China	1980	0.6	1999	0.4	−33.33
Indonesia	1973	1.1	1993	0.9	−18.18
Ethiopia	1977	1.4	2001–02	1	−28.57
Kenya			2005	0.86	
DRC	1970	1.5	1990	0.5	−66.67
Malawi			2011	0.72	
Tanzania			2009	1.5	

Source Compiled from FAO.

Africa have been experiencing fragmentation (Table 2). In one group of countries with small or marginal landholdings (less than 1 ha), farms have become smaller (China, India, Ethiopia, DRC and Indonesia) and in the other, medium landholdings have become small (Pakistan and Philippines). India is interesting because its average landholding size witnessed one of the highest percentage decreases. In the light of these trends, it becomes more important to assess the influences of small farm viability.

Understanding the relationship between land size and productivity is important to assess the potential and challenges for small farms, and to assess the impact of decreasing landholding sizes on growth of the agricultural sector. Landholding size and productivity have been debated in studies of rural development and economics for a long time. Since the 1960s, economists have argued that crop productivity per unit of land declined with an increase in farm size (Bardhan 1973; Mazumdar 1965; Sen 1962), which has led to the emergence of the 'small farm paradigm', which states that there is an inverse relationship between farm size and productivity. These studies conclude that small farms have an advantage over large farms in per capita productivity, due to higher labour utilisation (e.g. using family labour) and higher input utilisation (e.g. using intensive farming practices). This inverse relationship is a result of imperfect land and labour markets (Bardhan 1973; Sen 1966). Imperfections in the labour market meant that surplus labour was available at the household level, as off-farm opportunity costs (off-farm wages minus search and travel costs) were higher than on-farm wages, and low-cost labour allows for substituting 'lumpy' inputs such as capital-intensive equipment (Binswanger and Rosenzweig 1986; Eastwood et al. 2010; Poulton et al. 2010). Imperfect land markets meant that land lease options, to access more land for farming, were limited (Eswaran and Kotwal 1986; Hazell et al. 2010) and producers have had to effectively utilise their existing resource endowment. In some regions of South and Southeast Asia, landlords became credit providers to incentivise land lease and sharecropping (Basu 1997; Otsuka et al. 1992; Srivastava 1989). Therefore, in many Asian countries (where land was scarce and labour abundant), the 'small farm paradigm' did hold. In fact, this was considered a socially optimal outcome (Hazell et al. 2010; Poulton et al. 2010).

Since fixed costs are high on small farms, it is more difficult to take advantage of the economies of scale which can be beneficial to agricultural development. In fact, some studies show that the inverse relationship between small size and high productivity disappears when taking into account soil quality (Benjamin 1995; Bhalla and Roy 1988), capital market imperfections (Feder 1985) and unobserved heterogeneities such as climatic variations and quality of management (Eastwood et al. 2010). For example, capital market imperfections limit access to credit for farms with low land endowments, because they have limited value as collateral (Besley 1995a, 1995b; Bhaduri 1977; Ghosh 2013; Ghosh et al. 2001). This in turn constrains access to inputs, extension services, technology and lumpy inputs such as management and asset-specific machinery. Due to a limited volume of production, small farms often do not have bargaining power, which often leads to poor price realisation (Hazell et al. 2010; Johnson and Ruttan 1994; Poulton et al. 2010).

The development of smallholder agriculture is central to the structural transformation process in all developing countries. Growth in agricultural productivity leads to surplus creation and increased market participation by small farms, resulting in rising household-level incomes and welfare gains. This increased engagement with markets is referred to as commercialisation (Carletto et al. 2017; Pingali and Rosegrant 1995). Commercialisation is essential for the transfer of surplus in the form of food, labour and capital from the agrarian sector to the industrial and service sectors, to enable structural transformation (Timmer 1988). Different small farm-based economies are at various stages of structural transformation; they can be categorised as low-productivity agricultural systems, modernising agricultural systems and commercialised agricultural systems (Pingali et al. 2015). Figure 5 shows the performance of selected small farm-based economies at different stages of structural transformation. Countries with low per capita incomes and larger shares of agricultural contributions to GDP are referred to as low-productivity agricultural systems. Many of the SSA countries are classified as such and in these regions, hunger and poverty remain high.

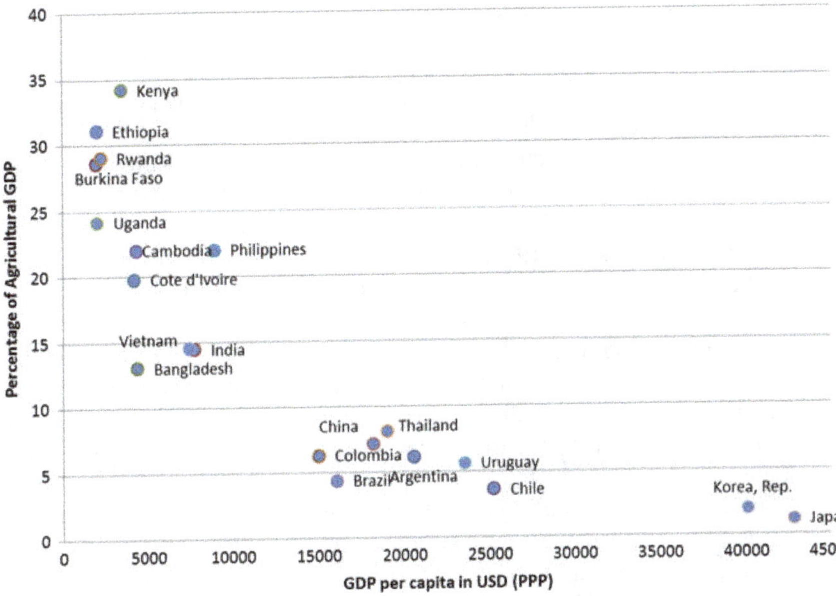

Fig. 5 Structural transformation and agricultural performance in selected countries. *Source* FAOSTAT and World Bank Data (2018)

Latin American, Southeast Asian and South Asian countries are classified as modernising agricultural systems, as they have medium-level per capita incomes between USD 5000 and 15,000 and their GDP contribution from agriculture is between 5 and 25%. These regions successfully implemented Green Revolution technologies and gained from the resulting agricultural productivity increases and have substantially reduced poverty and hunger. In these regions, however, there are high levels of income inequality and regional disparities in development. The small farm-dominated East Asian economies of Japan, Taiwan and South Korea are referred to as commercialised agricultural systems, as they have very high per capita incomes and low contributions to GDP from the agricultural sector. In the post-World War II period, these countries saw an increase in farm productivity, and surpluses were effectively transferred to other sectors to aid the structural transformation process. Small farms face different challenges in each of these production systems. By assessing the major characteristics of small farm economies at different stages of structural transformation, we can better understand the economic, nutritional, social and environmental challenges they face. This will enable us to evaluate the magnitude of the challenges to different economies in achieving the SDGs. In the following part of this section, we look at the major challenges faced by smallholder agriculture in each production system; the following section will assess the major interventions needed to remedy them.

3.1 *Low-Productivity Agricultural Systems*

Countries with low-productivity agricultural systems are beset with poor yields and incomes, despite having large land and/or labour inputs available. Most of these countries are in SSA, where there was low adoption of Green Revolution technologies in staple grains such as wheat, rice and maize (unlike in Asian and Latin American countries). While 82% of the area under staples in Asia comprised modern high-yielding varieties in 1998, in SSA this was only 27% (Evenson 2003). Figure 6 shows that the yield for cereals in Africa rose much less than it did in other regions of the world between 1961 and 2017. While cereal yield doubled in SSA, it quadrupled in South Asia, Latin America and Southeast Asia. The main reason is that agricultural production in low-productivity agricultural systems is carried out in marginal environments, with constraining agroclimatic, socio-economic and technological or biophysical constraints, where input-intensive Green Revolution technologies could not be adopted (Pingali et al. 2014). This is coupled with poor access to and provision of essential public goods such as R&D; factor markets such as credit, seeds, fertilisers and pesticides; and essential infrastructure such as irrigation, storage and roads; affecting production and incentives at the farm level. Development has also been affected by other challenges such as problematic governance, lack of institutional support (e.g. extension services and markets), and low and inelastic demand for agricultural products (Pingali 2010). In recent years, increases in productivity have occurred in these regions via area expansion, not through yield increases

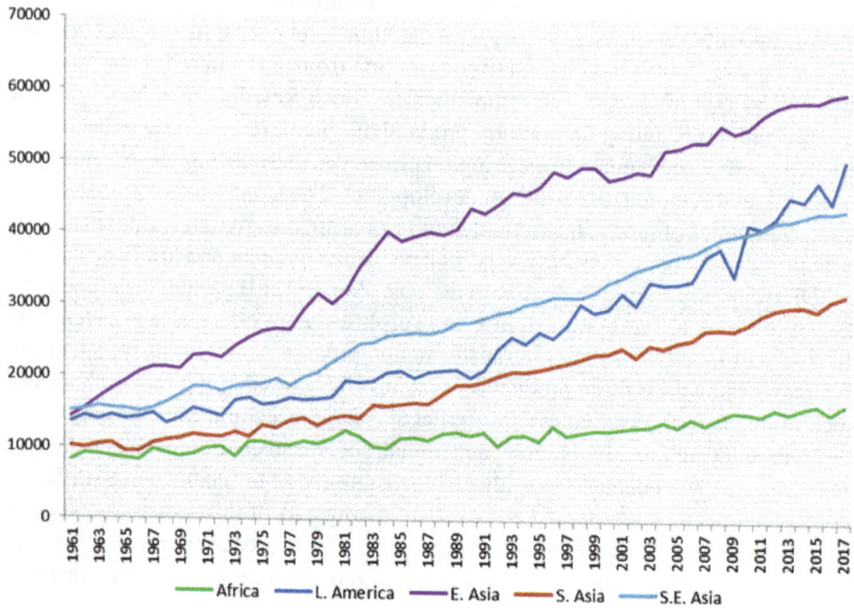

Fig. 6 Cereal yields for selected regions, 1961–2017 (hg/ha). *Source* FAOSTAT

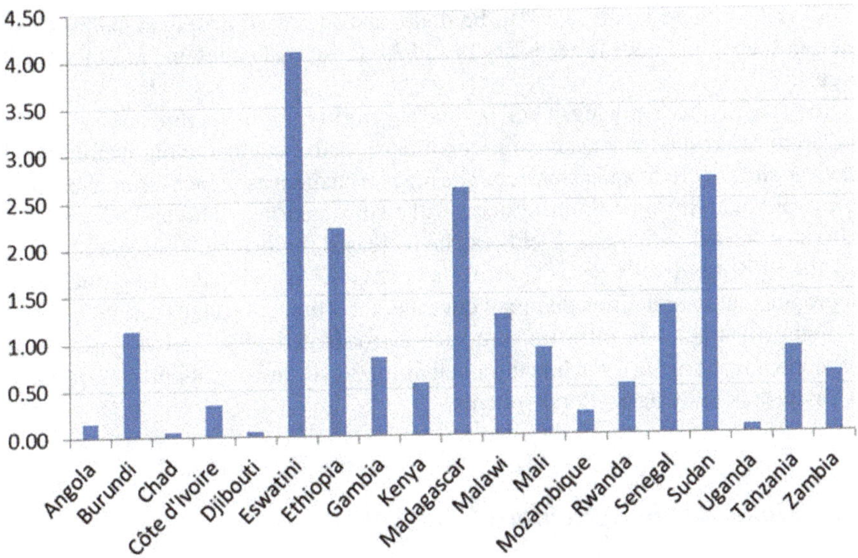

Fig. 7 Percentage of area under irrigation in selected net food-importing African nations, 2016. *Source* FAOSTAT

(Binswanger-Mkhize and McCalla 2010). Reforms since the 1990s have rectified incentive-distorting policies in agriculture in many countries (Anderson and Masters 2009), but growth and development in the agricultural sector remain challenging.

Environmental and climate change issues are among the biggest challenges in low-productivity agricultural systems. As it is, 43% of the African continent is dryland and is prone to extreme weather events and climate change (Cooper et al. 2008; UNDP 2009). According to FAO data, the average area under irrigation in net food-importing countries in SSA is around 1% (Fig. 7). Water stress and drought are often exacerbated by land degradation in sub-humid and semi-arid conditions to a greater degree than in purely arid conditions (Adhikari 2013); this only reinforces a higher level of land degradation and low agricultural productivity in SSA (Nkonya et al. 2008). Agroclimatic risks, and the absence of irrigation facilities and technological interventions such as drought-resistant crops, have resulted in high yield gaps in cereals and coarse grains, leading to calorie and micronutrient access problems.

Food and micronutrient access remain a major obstacle in low-productivity agricultural systems. These systems have high prevalence of child stunting, wasting and micronutrient deficiency, and these countries require a significant turnaround to accomplish the goals for hunger and poverty reduction (Pingali et al. 2015). According to FAO, IFAD and WFP (2015), 35.4% of the world's undernourished live in low-productivity agricultural systems in SSA. Here, despite a 44.4% drop in incidence since 1991, 19.3% of children under 5 were undernourished in 2015. The prevalence of stunting decreased from 49.0% in 1991 to 35.2% in 2015, while prevalence of wasting reduced from 11.0 to 8.2%. Noteworthy progress was made in the

MDG era but there is still work to be done, and increased access to calories and micronutrient-rich foods is necessary to address these challenges and meet the SDG targets.

To get agriculture moving, it will be vital to prioritise infrastructure investments in irrigation, watershed management programmes, roads, and marketing facilities and services such as credit and extension. The biggest challenges for achieving the SDGs in low-productivity agricultural systems will be in increasing yields and reducing the impact of agroclimatic risks. Yields can be increased through better access to R&D, credit and better-quality seeds from factor markets. Agroclimatic risks (including environmental externalities that may contribute to climate change) can be reduced through investments in infrastructure such as roads, irrigation and storage. These investments are crucial for achieving small farm growth and development, especially in low-productivity agricultural systems.

3.2 *Modernising Agricultural Systems*

Countries with modernising agricultural systems successfully implemented Green Revolution technologies and were able to reap the rewards of increased productivity of staple grains—poverty reduction, increased availability of food grains and lower food prices (Pingali 2012). In these economies, agricultural development also stimulated growth in non-agricultural sectors, resulting in rising incomes and urbanisation, which in turn led to rapid diversification of diets and boosted demand for higher-value crops and livestock products (Pingali 2007; Pingali and Khwaja 2004; Reardon et al. 2009; Reardon and Minten 2011). While this has created opportunities for small farms in terms of diversifying into high-value crops that could increase farm-level incomes, larger problems remain in access to key inputs such as credit and R&D, and to agricultural markets.

Despite the significant changes in nutritional status and access to food that have occurred in the past few decades, issues such as undernourishment, micronutrient deficiency, climatic challenges and interregional inequalities remain major constraints. Figure 8 shows that the prevalence of stunting in South Asia remains the highest globally, higher than in SSA. Despite a 47% drop since 1991, 32.7% of children under 5 are stunted. Although East Asia and Pacific, and Latin America and the Caribbean, have significantly reduced the prevalence of stunting, 67.7% and 61.1% (respectively) of children under 5 are stunted. Figure 8 also shows that the prevalence of overweight children under 5 is increasing in modernising agricultural systems. Latin America and the Caribbean, followed by East Asia and Pacific, have shown the highest growth in overweight, while South Asia has also witnessed a steady climb. In modernising agricultural systems, the problems of undernourishment, micronutrient deficiency and overnutrition therefore exist simultaneously.

Modernising agricultural systems also have significant regional and interregional differences that have led to unequal growth and development. Figure 9 shows the interregional differences in SSA and modernising agricultural systems. In Latin

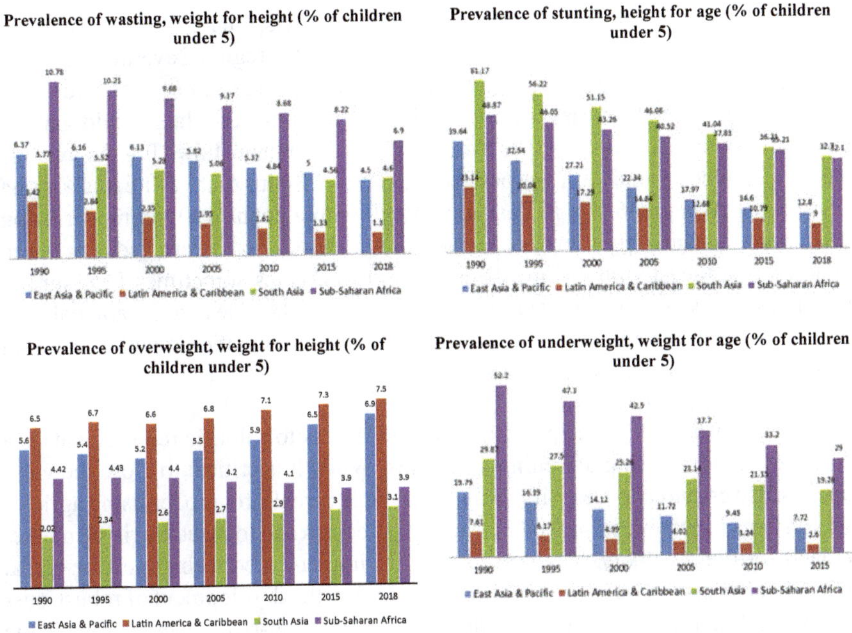

Fig. 8 Trends in wasting, stunting, overweight and underweight children in different regions world. *Source* FAOSTAT

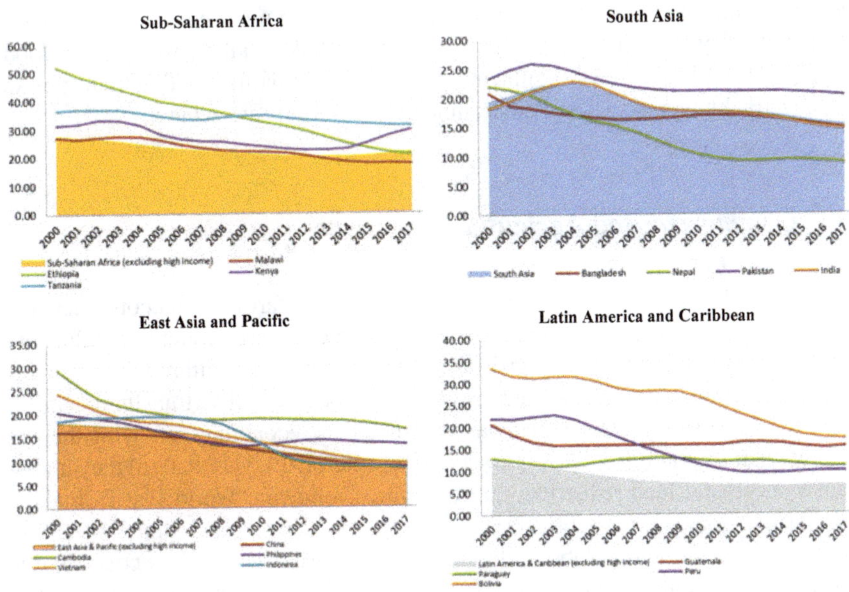

Fig. 9 Prevalence of undernourished (%) in selected regions and countries. *Source* FAOSTAT

America, for example, the small farm economies of Bolivia, Guatemala and Paraguay show higher prevalence of undernourishment compared to regional averages (Fig. 9). In terms of yield gap in cereals, Bolivia and Guatemala are seen to have over 50% less yield than other countries. Similarly, there are significantly large yield gaps in cereals in South Asia compared to other regions. FAO data show (Fig. 6) average cereal yield to be around 3 tonnes per hectare (t/ha) in South Asia, while it is 4.4 t/ha in Southeast Asia and 6 t/ha in East Asia. Even within countries with modernising agricultural systems, high levels of inequality exist, e.g. in eastern India, western China and northeast Brazil (Pingali 2010). These regions sometimes face similar challenges to low-productivity agricultural systems in SSA, due to geographical constraints such as poor market connectivity and low agroclimatic potential resulting from weather-related stress, e.g. droughts.

Undernourishment (or the prevalence of low access to calorific requirement) is a challenge in modernising agricultural systems. While it is critical to close the yield gap, it is also essential to diversify away from staple cereals, for modernising agricultural systems to achieve improved nutritional outcomes and to meet the rising market demand for diet diversity. Having successfully implemented techniques for cereal intensification using Green Revolution technologies, the development of robust, sustainable and market-oriented production of micronutrient-rich, diverse crops is the logical and necessary next step. To adapt to changing market demand for higher-value crops and other non-staples in ways that would benefit small farms, infrastructure and support resources would be required to enable them to participate in the value chain. The major challenges for modernising systems are in diversifying staple grain systems to improve access to calories, and in addressing micronutrient deficiencies and issues of overnutrition, in order to effectively link small farms to urban food value chains and to address significant interregional disparities within countries so that poverty, hunger, social and environmental SDG goals are achieved.

3.3 Commercialised Agricultural Systems

Taiwan, South Korea and Japan are examples of small farm-based economies that have low agricultural GDP and high per capita incomes. These economies underwent successful structural transformation by transitioning out of agriculture-based systems to manufacturing and service-based economies, through effective transfer of surplus out of the agricultural sector. What was unique to these economies, allowing them to develop their small farm sectors into successful enterprises, was that: (a) they underwent a successful land redistribution programme following World War II, making their farm sectors homogeneous (Francks et al. 1999; Ohkawa and Shinohara 1979); (b) these economies were labour-abundant and land-scarce, which is conducive to labour-intensive cultivation, at early stages of development when wages were low and cost of mechanisation was high (Hayami and Ruttan 1971); and (c) institutional arrangements such as cooperatives were set up to remedy problems faced by small farms in accessing inputs, credit, agricultural R&D and output markets (Huang 2006; Kajita 1965; Lin 2006).

In the past few decades, the profitability of small farms in commercialised agri-cultural systems has been declining, due to rising wages which have reduced the comparative advantage of labour-intensive farming systems (Otsuka et al. 2013). Consequently, assistance to the agricultural sector, in the form of subsidies and trade protection, has risen to keep agriculture artificially attractive (Anderson 2011). Con-solidation of landholdings to increase farm size is essential to improve the compara-tive advantage in these countries.

However, laws setting a ceiling on the landholding sizes of households prevent consolidation despite shrinking rural populations and continue to keep farms labour-intensive and small-scale (Otsuka et al. 2016). Mod-ernising agricultural economies, especially Indiaand China, mayalso follow asimilar trajectory, with rising rural wages. The key lessons for small farm production from these economies include the importance of improved access to factor and product markets, and the important role played by institutional arrangements such as coop-eratives in enabling these essential services. In the long run, rising wages in the agricultural sector may bring into question the comparative advantage of agricul-ture in modernising agricultural economies, and in the wake of this it may become essential to revisit the question of land consolidation.

The challenges to tackling hunger, poverty, environmental degradation and social disadvantages in accessing markets differ between small farm-dominated low pro-ductivity and modernising agricultural systems. In smallholder-dominated commer-cialised agricultural systems in Japan, South Korea and Taiwan, many of the disad-vantages to small farms were remedied through effective land reforms and institu-tional interventions in the form of cooperatives.

This enabled them to successfully transfer labour and capital resources from the agricultural sector to rapidly initiate structural transformation. In developing countries, improving the viability of small farms is central to the reduction of poverty and hunger and to the structural trans-formation process. Access to markets is important to incentivise production and diversification and to raise household-level incomes. However, market participation and commercialisation are conditional on transaction costs that influence access to essential factors of production and affect price realisation of produce and commodi-ties sold in the markets. The next section of this chapter assesses various transaction costs that influence access to markets in different production systems, and the inter-ventions that could rectify them to make smallholder farming more sustainable and viable.

4 Transaction Costs and the Commercialisation Process—Interventions and Policy Options in Different Production Systems

Agricultural markets in developing countries are complex institutions, incorporating many forms of production linkages and exchange relations (Benjamin et al. 2002; Harriss-White 1995). These markets comprise factor and product markets. From factor markets, agricultural producers access credit, technology, land and labour for production; in the product markets, they sell their produce. Increased market participation also marks the transition from subsistence-based agriculture to commercialised agriculture. Agricultural commercialisation takes place at the factor (input) and product (output) market levels (von Braun and Kennedy 1994). At the factor market level, commercialisation leads to progressive substitution of non-tradable inputs with tradable inputs used in agricultural production (Pingali and Rosegrant 1995); at the product market level, commercialisation means increased marketability of surplus and diversified products, which increases household-level incomes.

However, the ability to participate in these markets is determined by transaction costs, or the cost of accessing goods and services and making exchanges (Key et al. 2000). These costs limit the ability of smallholders to effectively participate in markets, hindering commercialisation. It is important to identify the characteristics of transaction costs, in order to determine their influence on smallholder production in different production systems, and to design interventions to address them. Smallholder systems are not homogeneous, as various farm, social and behavioural, geographical/locational and crop-specific characteristics influence the conditions under which production and marketing take place (Pingali et al. 2005). Table 3 highlights the main features of these characteristics in low productivity and modernising production systems, and the policy interventions required to address them. In low-productivity systems, high transaction costs of accessing factor markets are the major challenge that hinders productivity growth, resulting in low surpluses, low factor market participation and therefore low commercialisation. In modernising agricultural systems, the transaction costs of accessing product markets are high, and this influences smallholder ability to access specialised value chains and incentivise diversification of production, affecting incomes and growth. In this section, we look at how farm, household, location and crop-specific characteristics determine transaction costs in different production systems, and suggest policy interventions to address them. This is crucial to improve the productivity and incomes of smallholders, which is essential to achieving the SDGs.

Table 3 Transaction costs and interventions for different production systems

Transaction costs	Production system	Characteristics	Interventions
Farm-level	Low productivity	• Low yields and lack of marketable surplus, leading to 'low equilibrium poverty traps' • Low participation in factor markets due to high costs of accessing credit, inputs and R&D	• Improved access to capital markets (credit and insurance) • Provision of R&D as a public good along with extension services and supplementary inputs
	Modernising	• High entry cost for smallholders in high-value chains, limiting commercialisation • Low bargaining power in product markets	• Institutional interventions such as producer organisations • Market reforms • Public–private partnerships
Household-level	Low productivity	• High participation by women in agricultural labour force • High costs in accessing factor markets and productive resources for women	• Gender-focused policy initiatives • Provision of public goods, credit, R&D, specifically to women • Infrastructure to reduce women's household workload
	Modernising	• Access problems for women-led households, in product markets and value chains • Disproportionate workload on farms and at household level	• Gender-sensitive value chains and improved access to product markets • Investment in gender-focused labour-saving technologies • Women's groups to reduce costs related to scale and low bargaining power
Location-specific	Low productivity	• Vulnerable to climate change due to agroclimatic conditions • Productivity is low, and high production risks make access to credit and inputs difficult	• Sustainable agricultural intensification, irrigation infrastructure • Climate-focused R&D in crops • Combine production-enhancing activities with conservation
	Modernising	• Connectivity and distance to markets determines cost of marketing • Location determines ability to form contracts in high-value chains	• Infrastructure of connectivity and storage • Specialised cold chain for perishables

(continued)

Table 3 (continued)

Transaction costs	Production system	Characteristics	Interventions
Crop-specific	Low productivity	• High cost of adopting crops and livestock programmes hinders diversification and results in low commercialisation	• Aid commercialisation through policy deregulation • Investment in infrastructure, capital markets and R&D
	Modernising	• Policy favouring production and marketing of staple grains reduces comparative advantage of other crops, reducing incentives to diversify	• Crop-neutral agricultural policy • Infrastructure development to connect to high-value chains and bring a private sector response

4.1 Farm-Specific Costs and Interventions in Different Production Systems

The volume of marketable surplus generated by smallholders determines their ability to participate in both factor and product markets. Some farms are either subsistence or semi subsistence-based, as they produce only for household consumption or they have limited engagement with markets (either buying or selling). Other smallholders are more commercialised, as they both buy and sell in the markets (de Janvry et al. 1991). Farm-specific transaction costs are costs that influence smallholder participation in markets determined by the production status of farms. Small farms may not participate in, or may have difficulty accessing, both factor and product markets due to limited surplus creation.

In low-productivity systems, there is a prevalence of low equilibrium poverty traps, where low surplus leads to low market participation, resulting in low incentives to improve production, again leading to poor yields (Barrett 2008). Small landholdings with low yields have higher costs associated with accessing institutional credit, due to high production risks and low collateral. Low access to credit in turn hinders the ability to access quality inputs such as seeds with high yield potential, fertiliser and pesticides, and R&D that may reduce risks and uncertainties. The major policy agenda in low-productivity agricultural systems requires the reduction of transaction costs for accessing factor markets, beginning with capital (credit and insurance), followed by inputs and extension services that will enable crop intensification (increased output per unit of input). Improving capital market access will increase ability to access inputs and safeguard against production risks. Smallholder access to R&D was critical to the success of the Green Revolution. In low-productivity agricultural systems, similar initiatives are needed, along with public services such extension and information dissemination to increase adoption of yield-increasing technologies.

In modernising agricultural systems, transaction costs in product markets need to be reduced for better price realisation and improved farm incomes for smallholders.

In many of these economies, demand for high-value agricultural products has created opportunities for small farms to diversify production and realise better profits by participating in value chains. However, smallholder exclusion, due to transaction costs characterised by bureaucratic, monitoring and management costs, causes small farms to be discriminated against in favour of larger farms when forming contracts, limiting their participation (Dolan and Humphrey 2000; Hazell et al. 2010; Reardon and Berdegué 2002; Reardon et al. 2003; Swinnen and Maertens 2007). Policy interventions in modernising agricultural systems need to promote initiatives to streamline marketing chains and enable forward and backward linkages for smallholders, through contracts in high-value chains. Institutional interventions such as producer organisations and cooperatives have helped to provide inputs, reduce transaction costs and also form market linkages (Barrett et al. 2012; Bellemare 2012; Boselie, Henson and Weatherspoon 2003; Briones 2015; Reardon et al. 2009; Schipmann and Qaim 2010). Promotion of these institutions will help smallholders to mitigate some of the transaction costs associated with market entry, as it addresses problems associated with economies of scale. Incentives are also needed in these production systems to attract public–private partnership and to collaborate with civil society organisations to enable such linkages. Productivity and income growth, through increased market participation by smallholder farmers, are central to achieving the goals for poverty (SDG 1 and SDG 8) and nutrition (SDG 2 and SDG 3), and the social goal of reducing inequalities within and among countries (SDG 10).

4.2 *Household-Specific Costs and Interventions in Different Production Systems*

Behavioural and social characteristics influence household-level decision-making and smallholder ability to access factor and product markets (Pingali et al. 2007). The costs of market participation determined by social and behavioural characteristics are referred to as household-specific transaction costs. The behavioural characteristics that influence the cost of engaging with markets are the household's level of aversion to risk and uncertainty, entrepreneurial ability and technical ability (Barrett et al. 2012). In a more complex way, the social characteristics that influence market access and participation are social networks, caste (Sen 2000; Thorat 2009), age, gender (Agarwal 1995, 2010; RFST 2005) and education (Narayanan 2014). Here, we will look specifically at the issue of gender and its influence on transaction costs in smallholder agricultural production.

Women are among the largest groups of landless labourers, and the largest group dispossessed or with restricted access to land (Agarwal 1994; Deere and Leon 2001). They also represent two-thirds of livestock keepers (Thornton et al. 2002) and 30% of labour in fisheries (FAO 2011). Despite having an important role in production, studies have also shown women to face high costs in accessing capital, engaging in entrepreneurial activities (Fletschner and Carter 2008) and adopting technological inputs and mechanisation (FAO 2006). Therefore, in many developing countries, women-led households have lower yields and incomes, due to poor access to markets and productive resources (Croppenstedt et al. 2013), affecting their contributions to agricultural productivity (FAO 2011). Women also provide non-marketable goods and services at the household level, such as gathering water and fuel, child health and nutrition and also subsistence crop production which is essential for household welfare (Floro 1995). In this context, time-saving measures are relevant to women's workloads, income and household-level welfare.

In low-productivity agricultural systems, women's participation in the agricultural labour force is higher than the global average (Croppenstedt et al. 2013). Therefore, it is crucial to close the gender gap and address gender-specific transaction costs and constraints to agricultural production, to increase agricultural productivity and women's empowerment. Improving access to factors of production, such as cultivable land and institutional credit, is central to providing women with control over productive resources in agriculture. Better access to public goods such as tap water, and other private goods such as clean fuel for household use, helps to improve women's health, reduce drudgery and free up labour for more productive activities. Agricultural policies related to natural resource management, access to inputs and technology, and production affect male-headed households and female-headed household differently, and therefore, there is a need for a more gendered policy focus in agriculture (FAO 2011). It is essential to promote women's self-help groups (SHGs) for education, information dissemination, access to microcredit, provision of essential public goods and support for production-based activities. Investment is needed in infrastructure and capital for access to tap water and clean fuels for cooking, to free up time for more productive activities. Time-saving measures can also deliver multi-generational nutritional benefits to households with women using their freed time for other productive activities. This will also help to improve productivity and surplus creation in women-led household in low-productivity agricultural systems.

The two major interventions needed in modernising agricultural systems, to address gender-specific challenges, are improved access to product markets and labour savings for rural women. With regard to access to product markets, studies have shown that women involved in both traditional and modern crop production and marketing face considerable disadvantages and risks (Cabezas et al. 2007). A more gender-sensitive value chain is required to address access problems in markets (Rubin and Manfre 2014; Nakazibwe and Pelupessy 2014; Quisumbing et al. 2015). As women are often involved in agricultural labour and non-marketed household labour, measures to improve the labour efficiency and productivity of women will enable cost savings and free up time. Labour-saving technology needs to be implemented through mechanisation in agriculture, to reduce drudgery.

Policy initiatives to promote women's organisations, and build capacity to make them self-sustaining, is important to tackle gender-specific challenges in production and marketing. Gender-sensitive value chains that facilitate women's participation in high-value markets are essential. Supporting women's groups to form contracts, and building in support systems to enforce contracts and prevent hold-ups, is important to enable market linkages between farm and market, and to improve incomes through better price realisation. Collaboration with state and civil society organisations is vital to promote and empower women's producer organisations and SHGs. Mechanisation, like marketing, is scale-sensitive and collective action to enable joint access to labour-reducing machinery is again vital. Targeting of mechanisation in women-dominated activities in agriculture, such as transplantation and harvesting, needs to take precedence in modernising agricultural systems. It is important to address the household-specific transaction costs that influence women-led smallholder households, in improving productivity and agricultural growth to meet the poverty goals (SDG 1 and SDG 8). Improving time use and efficiency will play an important role in meeting the nutritional goals (SDG 2 and SDG 3) in different production systems. Economic empowerment of women is also central to meeting the social goals (SDG 5 and SDG 10).

4.3 Location-Specific Transaction Costs in Different Production Systems

Seasonality and geographical dispersion are major influences in agricultural production, making the location of farms an important determinant of agroclimatic risks, cost of production and marketing in different production systems. In low-productivity agricultural systems, climate change issues and environmental externalities are a pressing concern, as much of the cultivated area is unirrigated and/or semi-arid. In the long run, rising temperatures will affect yields and farm-level revenues in many of these regions (Kurukulasuriya et al. 2006). Mitigating the effects of climate change, and the need to simultaneously increase yield, will pose a major challenge to the growth and development of agricultural sectors in low-productivity agricultural systems. It is important to achieve agricultural intensification without increasing negative externalities of agricultural production, such as diminishing biodiversity, increasing GHG emissions, land and water degradation. Policy interventions to promote sustainable agricultural intensification are essential, to manage the dual challenge of climate change and productivity growth (Matson et al. 1997; Pretty et al. 2011). In the formulation of agricultural policy, it will prove to be essential to supplement agricultural productivity programmes with agroforestry for carbon sequestration, soil conservation and watershed management programmes to limit land degradation and promote water conservation (Lipper et al. 2006; Pretty et al. 2011). To offset the current impacts of climate change, investment is needed in R&D to promote heat and drought-resistant crops, technologies and infrastructure such as micro-irrigation systems. It is also crucial to make these technologies easily accessible to smallholders. These interventions are also relevant in modernising agricultural systems that have infrastructural disadvantages related to agroclimatic conditions. Therefore, farm-level transaction costs influenced by economies of scale, and household-level transaction costs determined by gender and social status, also need to be rectified to enable access to interventions to mitigate locational disadvantages.

In modernising agricultural systems, location-specific factors influence small farm linkages. In vertical coordination, agents are selective about farmers' eligibility to participate in contracts, depending on the location of farms (areas with good retail and processing infrastructure) (Martinez 2002a; Mishra and Chand 1995; Trebbin and Franz 2010). Procurement distance is also an important determinant in value chains of perishable products such as milk, poultry, eggs and meat (Martinez 1999, 2002b). In the context of the emerging relevance of organised retail, these factors determine whether an area has high or low potential for market linkages (Pingali et al. 2007). The preference by retailers for high potential areas can lead to market segmentation and exclusion of some farms (David and Kusterer 1990; Little and Watts 1994) and can increase regional disparity in modernising agricultural systems. Policy interventions to create infrastructural public goods, and mitigate locational disadvantages, in low potential areas will help to decrease regional disparity in market access. Increased investment by the state to expand storage facilities and cold chains and improve connectivity is also vital, to reduce wastage and increase marketing options for smallholders. These interventions in infrastructure are often needed for a private sector response to engage in markets and enable the emergence of vertical coordination where farms can directly connect with retail. Location-specific transaction costs can lead to higher degradation of land and natural resources—without proper management practices in these resources—and increase wastage in food products; it is vital to mitigate them for responsible and sustainable production and consumption. Therefore, addressing these costs, and factors that contribute to them, is central to achieving the environmental and poverty goals.

4.4 Crop-Specific Transaction Costs in Different Production Systems

The production and marketing of different crops have varying levels of transaction costs associated with them. The level of these costs and returns incentivises the adoption of crops at the farm level. Agricultural produce can be classified as commodities or products, where commodities are 'standardised agricultural products that have had little or no processing and often are raw materials for further procession' (Schaffner et al. 1998, p. 6). Grains and pulses are often considered to be commodities. Products are produce or subsets of a given commodity that is highly differentiated based on attributes (organic, processed, branded, variety, perishability) (Reardon and Timmer 2007). Fruits and vegetables, milk and dairy products, and meat are all considered products. Diversifying away from staples such as wheat, rice and maize, and towards higher-value crops is an integral part of commercialisation. This changes access to factor and product markets; commodities and products are influenced by different sets of production, marketing and transaction costs. Policy also plays an important role in influencing transaction costs. Subsidies and price support can lower production and marketing costs for certain crops, to distort incentives in their favour.

In low productive agricultural systems, diversification initially involves the addition of crops (coarse grains, micronutrient-dense legumes, pulses, vegetables and fruit) and livestock programmes to small farms, before moving towards specialised production (Pingali and Rosegrant 1995). A diversified production basket would enable better income opportunities, especially through livestock production and better access to nutritious food groups through the cultivation of coarse grains and pulses. Reducing transaction costs in accessing factor markets therefore becomes vital for diversification. Government policy plays a central role in aiding the commercialisation process, by developing capital markets to increase rural investments and improve access to credit and insurance; enabling access to R&D in income-enhancing technologies; continuing deregulation of the agricultural sector; and promoting health and nutrition-based initiatives to supplement the transition.

In modernising agricultural systems, diversification of production towards higher-value, market-oriented produce is needed for income growth. The major challenges for crop diversification have been both unfavourable policy and poor market access. In many countries that successfully implemented Green Revolution technologies, there is a policy bias favouring staple grains, especially wheat, rice and maize, through input subsidies and price support (Pingali 2015). Although they were initially incentives to adopt high-yielding technologies by lowering factor, production and marketing costs of staples, in the long run they have distorted farm and market-level incentives to diversify (ibid.). The costs of adopting more nutrition-rich or more commercially viable crops are relatively higher, due to subsidies for staples. In terms of market access, as products are highly differentiated, they require higher labour inputs, monitoring requirements, higher levels of credit, quality inputs and extension services. Due to higher transaction costs in accessing markets and forming contracts, there are low incentives for smallholders to commercialise.

Both policy and market interventions are required to rectify crop-specific transaction costs. At the policy level, crop-neutral agricultural policy is needed to level the playing field and enable better smallholder response to market and price signals (Pingali 2015). This entails the rollback of excessive support to staple grains in all production systems. Linking small producers to value chains is also important to enable fair price realisation and reduce market failures. It has been noted that small farms participating in value chains have both direct and indirect gains (Swinnen and Maertens 2007). The direct gains accrue through increase in productivity, improvement in quality and rise in household-level incomes (Birthal et al. 2009; Dries et al. 2009; Ramaswami et al. 2009). The indirect effects have been reduced risks in production, increased access to credit and technology, improved market participation and productivity spillovers to other crops (Bellemare 2012; Swinnen and Maertens 2007). Therefore, effective linkages to product markets play an important role in incentivising production, diversification and intensification in all production systems.

In this section, we looked at how transaction costs determined by farm, household, location and crop characteristics influence the ability of smallholders to access factor and product markets influencing livelihoods. These costs have varying influences on agricultural production and marketing in different farming systems, and therefore, policy and market interventions need to be tailored to specific challenges if various poverty, nutritional, social and environmental SDGs are to be met. In low-productivity agricultural systems, sustainable intensification of agriculture is needed to increase productivity, while keeping check on environmental externalities. The biggest challenge for these systems is the transaction costs associated with accessing factor markets. Policy interventions are needed to reduce the cost of accessing credit, quality inputs and R&D to support intensification and diversification, enabling income growth and nutritional outcomes. Technology enabling the production of high-yielding, heat and drought-resistant crops is also needed for sustainable production in the wake of rising temperatures and changing climate. In modernising production systems, with consumption demand changing to higher-value products, access to product markets determines the ability of smallholders to produce for these markets and improve household-level incomes. Therefore, it is also important to link small producers to value chains, and this requires investment in infrastructure for connectivity and storage. Institutional interventions, in the form of cooperatives and aggregation models, have worked well in rectifying smallholder disadvantages in commercialised agricultural systems. There is a need for more research and support in promoting these organisations more widely, as they have proven to reduce transaction costs and rectify smallholder disadvantages. In both production systems, women play an important role in agricultural production, and in producing non-marketable goods and services at the household level that increase welfare. Social emancipation and economic empowerment are important to improve income and nutritional outcomes and to meet the social goal of reduced gender inequality.

5 Conclusion

The 2030 Agenda for Sustainable Development approved by the UN member states set an ambitious goal to end poverty, protect the planet and ensure prosperity for all global citizens by the year 2030. The 17 SDGs have 169 targets designed to take a holistic approach to addressing the social, economic and environmental aspects of sustainable development. As the majority of global agricultural production takes place on small farms, and about 2 billion of the world's poor directly depend on the sector for their livelihood, working as cultivators or wage-earning labourers, the centrality of small farm development and growth to achieving the SDGs is undeniable. Nine of the 17 SDGs, pertaining to poverty eradication (SDG 1 and SDG 8), hunger and nutrition (SDG 1 and SDG 3), social emancipation and inequality (SDG 5 and SDG 10) and the environment (SDG 12, SDG 13 and SDG 15), are directly linked to the agricultural sector.

Small farms are heterogeneous and the production challenges they face are determined by their geography and the stage of structural transformation. Countries with low per capita incomes and a high share of GDP coming from agriculture are considered low-productivity agricultural systems, while countries with medium-level per capita incomes and less than 30% of GDP contribution from agriculture are considered modernising agricultural systems. Countries with high per capita income tend to be dominated by commercialised agricultural systems. In low productivity and modernising agricultural systems, productivity growth leading to surplus creation is essential to improve farm-level incomes and household-level welfare, to realise the SDGs. Achieving this requires increased market participation or commercialisation. However, market participation is determined by transaction costs and when these costs are high, commercialisation is hindered, affecting productivity and growth.

The process of agricultural commercialisation in developing countries is essential to meeting the poverty, nutritional, social and environmental SDGs. In this chapter, we use a transaction cost framework to assess the major costs that constrain small producer agriculture in different production systems, in order to identify specific areas of intervention needed to address them. In low-productivity agricultural systems, where yield increase is crucial to meet hunger and nutritional goals, improved access to factor markets is most important. Capital markets to access credit and insurance, R&D access to adopt high-yielding and climate change-resistant crops, and extension services to aid in diversification and effective utilisation of resources, are essential for increasing productivity while reducing environmental externalities. In modernising agricultural systems that have already witnessed productivity gains from the Green Revolution, access to product markets is essential. Here, the ability to access high-value chains, and form contacts with retailers and other end-users, is important to meet market opportunities and improve the incomes of smallholders. In both production systems, improving women's access to factor markets and product markets is also essential for productivity and household-level welfare. Time-saving measures are important to reduce drudgery for women in the production of both marketable and non-marketable goods and services. Improving access to clean water and fuel is important to save time at the household level, while promoting mechanisation would help to reduce labour time used in agricultural production.

Policy interventions are needed to rectify transaction costs and enable commercialisation. In low-productivity agricultural systems, policy is needed to reduce the cost of accessing credit, quality inputs and R&D to support intensification and diversification. In modernising agricultural systems, it is also important to rectify subsidies favouring wheat, rice and maize, to promote a crop-neutral agricultural policy, incentivising farm-level diversification towards other crops. A more gender-sensitive approach to agricultural policy is essential in all developing countries to address the social disadvantage women face in agricultural production and access to markets. Promoting aggregation models such as the cooperatives will also prove crucial, to rectify some of the scale disadvantages to smallholders in accessing markets. Gendered aggregated group such as SHGs will continue to play an important role in addressing gender-specific access problems, especially with capital markets and technology.

References

Adhikari, B. (2013). Poverty reduction through promoting alternative livelihoods: Implications for marginal drylands. *Journal of International Development, 25*(7), 947–967. https://doi.org/10.1002/jid.1820.

Agarwal, B. (1994). *A field of one's own: Gender and land rights in south Asia*. Cambridge: Cambridge University Press.

Agarwal, B. (1995). Gender and legal rights in agricultural land in India. *Economic and Political Weekly, 30*(12), A39–A56.

Agarwal, B. (2010). Rethinking agricultural production collectivities. *Economic and Political Weekly, 45*(9).

Anderson, K. (2011). Government distortions of agricultural prices: Lessons from rich and emerging economies. In K. Otsuka & K. Kalirajan (Eds.), *Community, market, and state*. Hampshire: Palgrave Macmillan.

Anderson, K., & Masters, W. A. (2009). *Distrotions to agricultural incentives in Africa*. Washington D.C.: The World Bank.

Bardhan, P. (1973). Size, productivity and returns to scale: An analysis of farm-level data in Indian agriculture. *Journal of Political Economy, 81*(6), 1370–1386.

Barrett, C. B. (2008). Smallholder market participation: Concepts and evidence from eastern and southern Africa. *Food Policy, 33*(4), 299–317. https://doi.org/10.1016/j.foodpol.2007.10.005.

Barrett, C. B., Bachke, M. E., Bellemare, M. F., Michelson, H. C., Narayanan, S., & Walker, T. F. (2012). Smallholder participation in contract farming: Comparative evidence from five countries. *World Development, 40*(4), 715–730. https://doi.org/10.1016/j.worlddev.2011.09.006.

Basu, S. (1997). Why institutional credit agencies are reluctant to lend to the rural poor: A theoretical analysis of the Indian rural credit market. *World Development, 25*(2), 267–280. https://doi.org/10.1016/S0305-750X(96)00103-9.

Bellemare, M. F. (2012). As you sow, so shall you reap: The welfare impacts of contract farming. *World Development, 40*(7), 1418–1434. https://doi.org/10.1016/j.worlddev.2011.12.008.

Benjamin, D. (1995). Can unobserved land quality explain the inverse productivity relationship? *Journal of Development Economics, 46*, 51–84.

Benjamin, D., Reardon, T., Stamoulis, K. G., & Winters, P. (2002). *Promoting farm/non-farm linkages for rural development—Case studies from Africa and Latin America*. Rome: Food and Agriculture Organisation, The United Nations.

Besley, T. (1995a). Property rights and investment incentives: Theory and evidence from Ghana. *Journal of Political Economy, 103*(5), 903–937.

Besley, T. (1995b). Savings, credit and insurance. In J. Behrman & T. N. Srinivasan (Eds.), *The handbook of developmental economics* (pp. 2124–2207). Amsterdam: Elsevier Science Publishers, B.V.

Bhaduri, A. (1977). On the formation of usurious interest rates in backward agriculture. *Cambridge Journal of Economics, 1*(4), 341–352. Retrieved from http://www.jstor.org/stable/23596381.

Bhalla, S. S., & Roy, P. (1988). Mis-specification in farm productivity analysis: The role of land quality. *Oxford Economic Papers, 40*(1), 55–73.

Binswanger-Mkhize, H., & McCalla., A. F. (2010). The changing context and prospects for agricultural and rural development in Africa. In P. L. Pingali & R. E. Evenson (Eds.), *Handbook of agricultural economics* (4th ed., Vol. 4, pp. 3571–3712). Oxford: Wlsevier B.V.

Binswanger, H. P., & Rosenzweig, M. R. (1986). Behavioural and material determinants of production relations in agriculture. *Journal of Development Studies, 22*, 503–539.

Birthal, P. S., Jha, A. K., Tiongco, M. M., & Narrod, C. (2009). Farm-level impacts of vertical coordination of the food supply chain: Evidence from contract farming of milk in India. *Indian Journal of Agricultural Economics, 64*(3), 481–496. Retrieved from http://search.proquest.com/docview/201483208?accountid=10267.

Boselie, D., Henson, S., & Weatherspoon, D. (2003). Supermarket procurement practices in developing countries: Redefining the roles of the public and private sectors. *American Journal of*

Agricultural Economics, *85*(5), 1155–1161. Retrieved from http://www.jstor.org.proxy.library.cornell.edu/stable/1244887.

Briones, R. M. (2015). Small farmers in high-value chains: Binding or relaxing constraints to inclusive growth? *World Development*, *72*, 43–52. https://doi.org/10.1016/j.worlddev.2015.01.005.

Byerlee, D., de Janvry, A., & Sadoulet, E. (2009). Agriculture for development: Toward a new paradigm. *Annual Review of Resource Economics, 1*, 15–31.

Cabezas, A., Reese, E., & Waller, M. (2007). *The wages of empire: Globalization, state transformation, and women's poverty* (A. Cabezas, E. Reese, & M. Waller, Eds.). Boulder CO: Paradigm Publishers.

Carletto, C., Corral, P., & Guelfi, A. (2017). Agricultural commercialization and nutrition revisited: Empirical evidence from three African countries. *Food Policy*, *67*, 106–118. https://doi.org/10.1016/j.foodpol.2016.09.020.

Carr, M., & Hartl, M. (2010). *Leightening the load: Labor saving technologies and practices for rural women.*

Chand, R., Prasanna, L. P. A., & Singh, A. (2011). Farm size and productivity: Understanding the strengths of smallholders and improving their livelihoods. *Economic and Political Weekly*, *46*(26&27), 5–11.

Christiaensen, L., Demery, L., & Kuhl, J.(2006). *The role of agriculture in poverty reduction an empirical perspective*. Policy Research Working Papers. The World Bank. http://doi.org/doi:10.1596/1813-9450-4013

Cooper, P. J. M., Dimes, J., Rao, K. P. C., Shapiro, B., Shiferaw, B., & Twomlow, S. (2008). Coping better with current climatic variability in the rain-fed farming systems of sub-Saharan Africa: An essential first step in adapting to future climate change? *Agriculture, Ecosystems & Environment*, *126*(1–2), 24–35. https://doi.org/10.1016/j.agee.2008.01.007.

Croppenstedt, A., Goldstein, M., & Rosas, N. (2013). Gender and agriculture: Inefficiencies, segregation, and low productivity traps. *World Bank Research Observer, 28*(1), 79–109. https://doi.org/10.1093/wbro/lks024.

David, G., & Kusterer, K. (1990). *Small farmer, big business-contract farming and rural development*. London: Macmillan.

de Janvry, A., Fafchamps, M., & Sadoulet, E. (1991). Peasant household behaviour with missing Markets: Some paradoxes Explained. *Economic Journal*, *101*(409).

Deere, C. D., & Leon, M. (2001). *Empowering women. Land and property rights in Latin America.* Pittsburgh: University of Pittsburgh Press.

Distefano, F. (2013). *Understanding and integrating gender issues into livestock projects and programmes: A checklist for practitioners*. Rome: Italy.

Dolan, C., & Humphrey, J. (2000). Governance and trade in fresh vegetables: The impact of UK supermarkets on the African horticulture industry. *Journal of Development Studies, 37*(2), 147–176.

Dries, L., Germenji, E., Noev, N., & Swinnen, J. F. M. (2009). Foreign direct investment, vertical integration, and local suppliers: Evidence from the Polish dairy sector. *World Development, 37*(11), 1742–1758.

Eastwood, R., Lipton, M., & Newell, A. (2010). Farm size. *Handbook of Agricultural Economics, 4,* 3323–3397. https://doi.org/10.1016/S1574-0072(09)04065-1.

Eswaran, M., & Kotwal, A. (1986). Access to capital and agrarian production organisation. *Economic Journal*, *96*(382), 482–498. Retrieved from http://onlinelibrary.wiley.com/journal/10.1111/%28ISSN%291468-0297/issues.

Evenson, R. E. (2003). Crop variety improvement and its effect on productivity: The impact of international agricultural research. In R. E. Evenson & D. Gollin (Eds.), (pp. 447–472). Cambridge, MA: CABI.

Fan, S., & Chan-Kang, C. (2005). Is small beautiful? Farm size, productivity, and poverty in Asian agriculture. *Agricultural Economics, 32,* 135–146. https://doi.org/10.1111/j.0169-5150.2004.00019.x.

Fan, S., Hazell, P., & Thorat, S. (2000). Government spending, growth and poverty in rural India. *American Journal of Agricultural Economics, 82*(4), 1038–1051. Retrieved from http://www.jstor.org/stable/1244540

FAO. (2006). *Agriculture, trade negotiations and gender*. Rome: Italy.

FAO. (2011). *The role of women in Agriculture* (ESA Working Paper No. 11–2). Retrieved from www.fao.org/economic/esa.

FAO. (2014). *State of food and agriculture*. Rome, Italy. Retrieved from http://www.fao.org/3/a-i4040e.pdf.

FAO, IFAD, & WFP. (2015). *The state of food insecurity in the world 2015. Meeting the 2015 international hunger targets: Taking stock of uneven progress*. Rome, Italy.

Feder, G. (1985). The relation between farm size and farm productivity. *Journal of Development Economics, 18,* 297–313.

Fletschner, D., & Carter, M. R. (2008). Constructing and reconstructing gender: Reference group effects and women's demand for entrepreneurial capital. *The Journal of Socio-Economics, 37*(2), 672–693. http://doi.org/10.1016/j.socec.2006.12.054.

Floro, M. S. (1995). Economic restructuring, gender and the allocation of time. *World Development, 23*(11), 1913–1929. https://doi.org/10.1016/0305-750X(95)00092-Q.

Fontana, M. (2009). The gender effects of trade liberalisation in developing countries: A review of the literature. In M. Bussolo & R. De Hoyos (Eds.), *Gender aspects of the trade and poverty nexus: A micro-macro approach*. Palgrave Macmilan: Baskingstoke.

Francks, P., Boestel, J., & Kim, C. H. (1999). *Agriculture and economic development in east Asia-from growth to protectionism in Japan, Korea and Taiwan*. London: Routledge.

Ghosh, J. (2013). Microfinance and the challenge of financial inclusion for development. *Cambridge Journal of Economics, 37*(6), 1203–1219.

Ghosh, P., Mookherjee, D., & Ray, D. (2001). Credit rationing in developing countries: An overview of the theory. In D. Mookherjee & D. Ray (Eds.), *Readings in the theory of economic development* (pp. 283–301). Malden, MA: Blackwell Publishing Ltd.

Goldstein, M., & Udry, C. (2005). *The profits of power: Land rights and agricultural investment in Ghana* (Economic Growth Center Working Paper No. 929). New Haven, CT.

Harriss-White, B. (1995). Maps and landscapes of grain markets in South Asia. In J. Harriss, J. Hunter, & C. M. Lewis (Eds.), *The new institutional economics and Third World development.* London ; New York: Routledge.

Hayami, Y., & Ruttan, V. W. (1971). *Agricultural development: An international perspective.* Baltimore: Johns Hopkins University Press.

Hazell, P., Poulton, C., Wiggins, S., & Dorward, A. (2010). The future of small farms: Trajectories and policy priorities. *World Development, 38*(10), 1349–1361. https://doi.org/10.1016/j.worlddev.2009.06.012.

Huang, J. (2006). *Taiwan in transformation, 1895–2005: The challenge of a new democracy to an old civilization*. New Jersey: Transaction Publishers.

Ivanic, M., & Martin, W. (2008). Implications of higher global food prices for poverty in low-income countries1. *Agricultural Economics, 39,* 405–416. https://doi.org/10.1111/j.1574-0862.2008.00347.x.

Johnson, N. L., & Ruttan, V. W. (1994). Why are farms so small? *World Development, 22*(5), 691–706. https://doi.org/10.1016/0305-750X(94)90044-2.

Jost, C., Kyazze, F., Naab, J., Neelormi, S., Kinyangi, J., Zougmore, R., … Kristjanson, P. (2016). Understanding gender dimensions of agriculture and climate change in smallholder farming communities. *Climate and Development, 8*(2), 133–44.

Kajita, M. (1965). Land policy after land reforms in Japan. *Developing Economies, 3*(1), 88–105. https://doi.org/10.1111/j.1746-1049.1965.tb00749.x.

Key, N., Sadoulet, E., & de Janvry, A. (2000). Transactions costs and agricultural household supply response. *American Journal of Agricultural Economics, 82*(2), 245–259. https://doi.org/10.1111/0002-9092.00022.

Kraemer, K., & Zimmermann, M. (2007). *The guidebook nutritional anemia*. Basel, Switzerland: Sight and Life Press.

Kurukulasuriya, P., Mendelsohn, R., Hassan, R., Benhin, J., Deressa, T., Diop, M., … Dinar, A. (2006). Will African agriculture survive climate change? *The World Bank Economic Review, 20*(3), 367–388. Retrieved from https://doi.org/10.1093/wber/lhl004.

Leichenko, R. M., & O'Brien, K. L. (2002). The dynamics of rural vulnerability to global change: The case of southern Africa. *Mitigation and Adaptation Strategies for Global Change, 7*(1), 1–18. https://doi.org/10.1023/A:1015860421954.

Lin, D. (2006). Agricultural cooperatives in Taiwan. In *2006 FFTC-NACF International Seminar on Agricultural Cooperatives in Asia: Innovations and Opportunities in the 21st Century*. Seoul, Korea. Retrieved from http://www.agnet.org/htmlarea_file/activities/201.

Lipper, L., Pingali, P., & Zurek, M. (2006). Less-favoured areas: looking beyond agriculture towards ecosystem services. In R. Ruben, J. Pender, & A. Kuyvenhoven (Eds.), *Sustainable poverty reduction in less-favoured areas: Problems, options and strategies* (pp. 442–460). Wallingford, UK: CABI.

Little, P., & Watts, M. (1994). *Living under contract: Contract farming and agrarian transformation in Sub-Saharan Africa*. Madison, WI: University of Wisconsin Press.

Martinez, S. W. (1999). *Vertical coordination in the pork and broiler industry* (AER No. 777). Washington DC.

Martinez, S. W. (2002a). *A comparison of vertical coordination in the US poultry, egg, and pork industries* (Agriculture Information Bulletin No. 747–5). Washington DC.

Martinez, S. W. (2002b). *Vertical coordination of Marketing systems: Lessons from the poultry, egg and pork industries* (Agricultural Economic Report No. 807).

Matson, P. A., Parton, W. J., Power, A. G., & Swift, M. J. (1997). Agricultural intensification and ecosystem properties. *Science, 277*(5325), 504–509. https://doi.org/10.1126/science.277.5325.504.

Mazumdar, D. (1965). Size of farm and productivity—A problem of indian peasant agriculture. *Economica, 32*(126), 161–173.

Mishra, S., & Chand, R. (1995). Public and private capital formation in Indian agriculture: Comments on complementarity hypothesis and others. *Economic and Political Weekly, 30*(25), A64–A79.

Morton, J. F. (2007). The impact of climate change on smallholder and subsistence agriculture. In *Proceedings of the National Academy of Sciences, 104*(50), 19680–19685. Retrieved from http://www.pnas.org/content/104/50/19680.abstract.

Myers, S. S., Zanobetti, A., Kloog, I., Huybers, P., Leakey, A. D. B., Bloom, A. J., … Usui, Y. (2014). Increasing CO_2 threatens human nutrition. *Nature, 510*(7503), 139–142. Retrieved from https://doi.org/10.1038/nature13179.

Nagayets, O. (2005). Small farms: Current status and key trends. In *Information brief prepared for the "Future of Small Farms Research Workshop."* Wye College, Kent.

Nakazibwe, P., & Pelupessy, W. (2014). Towards a gendered agro-commodity approach. *Journal of World—Systems Research, 20*(2), 229–256. Retrieved from https://search.proquest.com/docview/1561499952?accountid=10267

Narayanan, S. (2014). Profits from participation in high value agriculture: Evidence of heterogeneous benefits in contract farming schemes in Southern India. *Food Policy, 44*, 142–157. https://doi.org/10.1016/j.foodpol.2013.10.010.

Nkonya, E., Pender, J., Kaizzi, K., Kato, E., Mugarura, S., Ssali, H., & Muwonge, J. (2008). *Linkages between land management, land degradation, and poverty in sub-Saharan Africa: The case of Uganda*. Washington D.C.

O'Brien, K., Leichenko, R., Kelkar, U., Venema, H., Aandahl, G., Tompkins, H., … West, J. (2004). Mapping vulnerability to multiple stressors: Climate change and globalization in India. *Global Environmental Change, 14*(4), 303–313. http://doi.org/10.1016/j.gloenvcha.2004.01.001.

Ohkawa, K., & Shinohara, M. (1979). *Patterns of Japanese economic development: A quantitative appraisal*. New Haven: Yale University Press.

Otsuka, K., Chuma, H., & Hayami, Y. (1992). Land and labor contracts in agrarian economies: Theories and facts. *Journal of Economic Literature, 30*(4), 1965–2018. Retrieved from http://www.jstor.org/stable/2727971.

Otsuka, K., Liu, Y., & Yamauchi, F. (2013). Factor Endowments, Wage Growth, and Changing Food Self-Sufficiency: Evidence from Country-Level Panel Data. *American Journal of Agricultural Economics, 95*(5), 1252–1258. Retrieved from https://doi.org/10.1093/ajae/aat028.

Otsuka, K., Liu, Y., & Yamauchi, F. (2016). The future of small farms in Asia. *Development Policy Review, 34*(3), 441–461. https://doi.org/10.1111/dpr.12159.

Paerl, H. W., & Huisman, J. (2009). Climate change: A catalyst for global expansion of harmful cyanobacterial blooms. *Environmental Microbiology Reports, 1*(1), 27–37. https://doi.org/10.1111/j.1758-2229.2008.00004.x.

Paterson, R. R. M., & Lima, N. (2010). How will climate change affect mycotoxins in food? *Food Research International, 43*(7), 1902–1914. http://doi.org/10.1016/j.foodres.2009.07.010.

Perez, C., Jones, E. M., Kristjanson, P., Cramer, L., Thornton, P. K., Förch, W., & Barahona, C. (2015). How resilient are farming households and communities to a changing climate in Africa? A gender-based perspective. *Global Environmental Change, 34*, 95–107. http://doi.org/10.1016/j.gloenvcha.2015.06.003.

Pingali, P. (2007). Agricultural growth and economic development: A view through the globalization lens. *Agricultural Economics, 37,* 1–12. https://doi.org/10.1111/j.1574-0862.2007.00231.x.

Pingali, P. (2010). Agriculture renaissance: Making "agriculture for development" work in the 21st century (Chap. 74). In P. Pingali & R. Evenson (Eds.), *Handbook of agricultural economics*, (Vol. 4, pp. 3867–3894). Elsevier. https://doi.org/10.1016/S1574-0072(09)04074-2.

Pingali, P. (2012). Green revolution: Impacts, limits, and the path ahead. *Proceedings of the National Academy of Sciences, 109*(31), 12302–12308. https://doi.org/10.1073/pnas.0912953109.

Pingali, P. (2015). Agricultural policy and nutrition outcomes—Getting beyond the preoccupation with staple grains. *Food Security, 7*(3), 583–591. https://doi.org/10.1007/s12571-015-0461-x.

Pingali, P. (2016). The hunger metric mirage: There's been less progress on hunger reduction than it appears. *Proceedings of the National Academy of Sciences of the United States of America, 113*(18).

Pingali, P., & Khwaja, Y. (2004). *Globalisation of Indian diets and the transformation of food supply systems.* (ESA Working Paper No. 04–05). Rome, Italy.

Pingali, P., Khwaja, Y., & Madelon, M. (2005). *Commercializing small farms: Reducing transaction Cost.* (ESA Working Paper No, 05–08).

Pingali, P., Khwaja, Y., & Madelon, M. (2007). The role of the public and private sectors in commercializing small farms and reducing transaction costs. In J. F. M. Swinnen (Ed.), *Global supply chains, standards and the poor* (pp. 260–267). Oxford, UK: CABI International.

Pingali, P., Ricketts, K., & Sahn, D. E. (2015). The fight against hunger and malnutrition: The role of food, agriculture, and targeted policies. In D. E. Sahn (Ed.), *Agriculture for nutrition*. Oxford, UK: Oxford University Press.

Pingali, P., & Rosegrant, M. W. (1995). Agricultural commercialization and diversification: Processes and policies. *Food Policy, 20*(3), 171–185. https://doi.org/10.1016/0306-9192(95)00012-4.

Pingali, P., Schneider, K., & Zurek, M. (2014). Poverty, agriculture and the environment: The case of sub-Saharan Africa. In J. von Braun & F. W. Gatzweiler (Eds.), *Marginality: Addressing the nexus of poverty, exclusion and ecology* (pp. 151–168). Dordrecht Heidelberg New York London: Springer.

Pingali, P., Witwer, M., & Abraham, M. (2016). Getting to zero hunger: Learnin from the MDGs for the SDGs. In P. D. Y. Lalaguna, C. M. D. Barrado, & C. R. F. Liesa (Eds.), *International society and sustainable development goals* (pp. 175–202). Pamplona, Spain: Thomson Reuters Aranzadi.

Poulton, C., Dorward, A., & Kydd, J. (2010). The future of small farms: New directions for services, institutions, and intermediation. *World Development, 38*(10), 1413–1428. https://doi.org/10.1016/j.worlddev.2009.06.009.

Pretty, J., Toulmin, C., & Williams, S. (2011). Sustainable intensification in African agriculture. *International Journal of Agricultural Sustainability, 9*(1), 5–24. https://doi.org/10.3763/ijas.2010.0583.

Quisumbing, A. R., & Pandolfelli, L. (2010). Promising approaches to address the needs of poor female farmers: Resources, constraints, and interventions. *World Development, 38*(4), 581–592. https://doi.org/10.1016/j.worlddev.2009.10.006.

Quisumbing, A. R., Rubin, D., Manfre, C., Waithanji, E., van den Bold, M., Olney, D., … Meinzen-dick, R. (2015). Gender, assets, and market-oriented agriculture: Learning from high-value crop and livestock projects in Africa and Asia. *Agriculture and Human Values, 32*(4), 705–725. https://doi.org/10.1007/s10460-015-9587-x.

Ramaswami, B., Birthal, P. S., & Joshi, P. K. (2009). Grower heterogeneity and the gains from contract farming: The case of Indian poultry. *Indian Growth and Development Review, 2*(1), 56–74. https://doi.org/10.1108/17538250910953462.

Rapsomanikis, G. (2015). *The economic lives of smallholder farmers- An analysis based on household data from nine countries*. Rome: Italy.

Ravallion, M., & Chen, S. (2007). China's (uneven) progress against poverty. *Journal of Development Economics, 82*(1), 1–42. http://doi.org/10.1016/j.jdeveco.2005.07.003.

Reardon, T., Barrett, C. B., Berdegué, J. A., & Swinnen, J. F. M. (2009). Agrifood industry transformation and small farmers in developing countries. *World Development, 37*(11), 1717–1727. https://doi.org/10.1016/j.worlddev.2008.08.023.

Reardon, T., & Berdegué, J. A. (2002). The rapid rise of supermarkets in Latin America: Challenges and opportunities for development. *Development Policy Review, 20*(4), 371–388.

Reardon, T., & Minten, B. (2011). Surprised by supermarkets: Diffusion of modern food retail in India. *Journal of Agribusiness in Developing and Emerging Economies, 1*(2), 134–161. https://doi.org/10.1108/20440831111167155.

Reardon, T., & Timmer, P. C. (2007). Transformation of markets for agricultural output in developing countries since 1950: How has thinking changed? (Chap. 55) In *Handbook of agricultural economics* (Vol. 3, pp. 2807–2855). Elsevier. https://doi.org/10.1016/S1574-0072(06)03055-6.

Reardon, T., Timmer, P. C., Barrett, C. B., & Berdegué, J. A. (2003). The rise of supermarkets in Africa, Asia, and Latin America. *American Journal of Agricultural Economics, 85*(5), 1140–1146.

RFST. (2005). *Impact of WTO on Women in Agriculture*. New Delhi.

Rojas-Downing, M. M., Nejadhashemi, A. P., Harrigan, T., & Woznicki, S. A. (2017). Climate change and livestock: Impacts, adaptation, and mitigation. *Climate Risk Management, Forthcomin*. http://doi.org/10.1016/j.crm.2017.02.001.

Rubin, D., & Manfre, C. (2014). Promoting gender-equitable agricultural value chains: Issues, opportunities, and next steps. In A. R. Quisumbing, R. Meinzen-Dick, T. Raney, A. Croppenstedt, J. A. Behrman, & A. Peterman (Eds.), *Gender in agriculture and food security: Closing the knowledge gap*. New York: Springer and FAO.

Schaffner, David, J., Schroder, W. R., & Earle, M. D. (1998). *Food marketing: An international perspective*. Boston, MA: WCB McGraw-Hill.

Schipmann, C., & Qaim, M. (2010). Spillovers from modern supply chains to traditional markets: Product innovation and adoption by smallholders. *Agricultural Economics, 41*(3–4), 361–371. https://doi.org/10.1111/j.1574-0862.2010.00438.x.

Sen, A. K. (1962). An aspect of Indian agriculture. *The Economic Weekly, 14*(46), 243–46.

Sen, A. K. (1966). Peasants and Dualism with or without Surplus Labor. *Journal of Political Economy, 74*, 425–450.

Sen, A. K. (2000). *Social exclusion: Concept, application, and scrutiny* (Social Development Paper No. 1). Bangkok.

Sheahan, M., & Barrett, C. B. (2014). *Understanding the agricultural input landscape in Sub-Saharan Africa recent plot, household, and community-level evidence* (Policy Research Working Paper No. 7014).

Smith, P., Martino, D., Cai, Z., Gwary, D., Janzen, H., Kumar, P., … Towprayoon, S. (2007). Agriculture. In B. Metz, O. R. Davidson, P. R. Bosch, R. Dave, & L. A. Meyer (Eds.), *Climate change 2007: Mitigation. Contribution of working group III to the fourth assessment report of the intergovernmental panel on climate change.* Cambridge University Press, Cambridge.

Srivastava, R. (1989). Interlinked modes of exploitation in Indian agriculture during transition: A case study. *The Journal of Peasant Studies, 16*(4), 493–522. https://doi.org/10.1080/03066158908438404.

Swinnen, J. F. M., & Maertens, M. (2007). Globalization, privatization, and vertical coordination in food value chains in developing and transition countries. *Agricultural Economics, 37,* 89–102. https://doi.org/10.1111/j.1574-0862.2007.00237.x.

Swinnen, J., & Squicciarini, P. (2012). Mixed messages on prices and food security. *Science, 335*(6067), 405–406. https://doi.org/10.1126/science.1210806.

Thirtle, C., Lin, L., & Piesse, J. (2003). The impact of research-led agricultural productivity growth on poverty reduction in Africa, Asia and Latin America. *World Development, 31*(12), 1959–1975. http://doi.org/10.1016/j.worlddev.2003.07.001.

Thorat, S. (2009). Economic exclusion and poverty linkages: A reflection on concept, consequences, and remedies in an Asian context. In J. von Braun, R. V. Hill, & R. Pandya-Lorch (Eds.), *The poorest and hungry assessments, analyses, and actions.* Washington D.C: International Food Policy Research Institute.

Thornton, P. K., Kruska, R. L., Henninger, N., Kristjanson, P. M., Reid, R. S., Atieno, F., … Ndegwa, T. (2002). *Mapping poverty and livestock in the developing world.* Nairobi, Kenya.

Timmer, P. C. (1988). The agricultural transformation (Chap. 8). In *Handbook of development economics* (Vol. 1, pp. 275–331). http://doi.org/10.1016/S1573-4471(88)01011-3.

Timmer, P. C., & Akkus, S. (2008). *The structural transformation as a pathway out of poverty: Analytics, empirics and politics* (Working Paper No. 150). Washington D.C.

Trebbin, A., & Franz, M. (2010). Exclusivity of private governance structures in agrofood networks: Bayer and the food retailing and processing sector in India. *Environment and Planning, 42,* 2043–2057.

Tubiello, F. N., Salvatore, M., Rossi, S., Ferrara, A., Fitton, N., & Smith, P. (2013). The FAOSTAT database of greenhouse gas emissions from agriculture. *Environmental Research Letters, 8*(1), 15009. Retrieved from http://stacks.iop.org/1748-9326/8/i=1/a=015009.

Tzannatos, Z. (1999). Women and labor market changes in the global economy: Growth helps, inequalities hurt and public policy matters. *World Development, 27*(3), 551–569. http://doi.org/10.1016/S0305-750X(98)00156-9.

UNDP. (2009). *Climate change in the African drylands: options and opportunities for adaptation and mitigation.* New York.

United Nations. (2015). *Transforming our world: The 2030 agenda for sustainable development.* Resolution adopted by the General Assembly on 25 September 2015 (A/RES/70/1). New York: United Nations General Assembly.

von Braun, J., & Kennedy, E. (1994). *Agricultural Commercialization, economic development and nutrient.* Baltimore: Maryland, USA: Johns Hopkins University Press.

Webb, P., & Block, S. (2012). Support for agriculture during economic transformation: Impacts on poverty and undernutrition. *Proceedings of the National Academy of Sciences, 109*(31), 12309–12314. Retrieved from http://www.pnas.org/content/109/31/12309.abstract.

9

Rural Development Strategies and Africa's Small Farms

Donald F. Larson, Rie Muraoka and Keijiro Otsuka

Abstract Improving the productivity of smallholder farms in sub-Saharan Africa (SSA) offers the best chance of reducing poverty among this generation of rural poor, by building on the few resources farming households already own. It is also the best and shortest path to meet rising food needs. Using examples from farmers' maize and rice fields, comparisons with Asia, and an extensive literature review, we explain why the set of technologies promoted to date have produced localised successes rather than transformational change. We also examine the limitations of alternative policies that are not centred on small farms. We give indicative examples of how resource management technologies can supplement seed-fertiliser technologies to speed an African Green Revolution.

1 Introduction

The goal of boosting productivity on smallholder farms is a central pillar in the rural development strategies of most African governments. There are many reasons for the broad support given to African smallholders, but two are most often cited. First, the vast natural resources in the hands of smallholder farmers in SSA can be used more productively to feed a growing global population, many of whom will live in Africa. Second, increasing the agricultural incomes through improved technologies offers the shortest path to poverty reduction in rural areas, where poverty has been most persistent. In this essay, we argue that, while it is the second argument that is

This chapter is derived from an original article written by the authors in Global Food Security in 2016. More details can be found in the acknowledgments of the chapter.

D. F. Larson (✉)
Global Research Institution, College of William and Mary, Williamsburg, USA
e-mail: don.larson@iides.org

R. Muraoka
Graduate Institute for Policy Studies, Japan Society for the Promotion of Science, Tokyo, Japan

K. Otsuka
Graduate School of Economics, Kobe University, Kobe, Japan

especially compelling for policy, finding ways to boost smallholder productivity in Africa offers the best chance of achieving both objectives.

On average, smallholder farmers in Africa would earn higher incomes working in sectors other than agriculture. Also, in countries where land and labour productivity are highest, farms are usually larger than they are in Africa. Furthermore, as economies develop, the proportion of workers in agriculture declines and a larger share of the population lives in cities. So, why should rural development strategies in SSA remain focused on smallholder farms?

The answer has to do with the slow pace of this archetypical economic transformation. For many reasons, the reallocation of labour from agriculture to other sectors is constrained and occurs over generations, even when income gaps are large and sustained. The pace of farm restructuring is slower still, so farms tend to remain small even as agriculture's share of employment declines. Consequently, policies designed to reduce poverty for this generation of rural poor must work largely within the constraints of small farms. In the case of Asia's Green Revolution, new technologies were developed by scientific institutions and quickly adopted by farmers that did just that. What's more, productivity growth rates were sustained while farms remained small. Ultimately, Asia's success influenced African policies, where most strategies to boost productivity have three common elements: a focus on smallholders; an emphasis on staple crops, mostly maize and rice; and a reliance on improved technologies, most often based on fertiliser-responsive high-yielding seeds.

Nevertheless, while the need to drive rural development through improved smallholder productivity is clear, the task is harder in SSA than it was in Asia. Specifically, the agroclimatic and market conditions that predicate the success of technologies are more varied in Africa than in Asia at the start of its Green Revolution. Consequently, a wider portfolio of technologies is needed to launch a transformational African Green Revolution and the task of identifying what works best locally is more difficult. It also means that the seed-fertiliser-focused technologies that work well in Asia, though important, are unlikely to solve all of the key constraints faced by African farmers. This, in turn, holds back sector-wide productivity growth and explains why successes to date have been local rather than sectoral. Part of the solution is to design and deploy a larger set of targeted technologies with local constraints in mind. Drawing on a set of farm studies originally published in Otsuka and Larson (2016), we document how resource management-based technologies can supplement conventional high-yielding seed-fertiliser technologies to improve smallholder productivity in SSA.

2 Green Revolutions as a Path Out of Rural Poverty

There are several potential pathways out of poverty for rural households, although none is easy. Family members from poor households often leave rural areas, migrating to cities or to other countries to earn incomes outside of agriculture. Still, studies suggest that potential migrants are often hampered by mismatched skills and anchored by illiquid land assets and place-specific social capital, which provide informal forms of insurance otherwise unavailable (Larson et al. 2004). Additionally, the benefits of moving away from agriculture decline with age. Consequently, the window for sectoral migration is brief and constrained. As a result, structural transformation is exceedingly slow and takes generations to achieve (Larson and Mundlak 1997; Gardner 2000; Butzer et al. 2003). In many African countries, it is a process that is far from complete. According to the Food and Agriculture Organization (FAO 2017), and due to the importance of the rural population of SSA (593 million in 2015), a continuous densification of the rural areas is taking place, and the absolute number of people living in rural areas will continue to climb until 2050 with a projected rural population of 909 million (United Nations 2018). In 2010, nearly 60% of jobs in SSA were in agriculture, and more jobs are expected to be added in this sector by 2020 than in the formal service and industry sectors (Fox et al. 2013).[1] Consequently, a large portion of the rural poor will remain in agriculture for the foreseeable future and any effective set of policies will have to reach them there.

Rural non-farm income activities offer another path out of poverty and can be key to achieving food security (Otsuka and Yamano 2006; Dethier and Effenberger 2012). However, agriculture is often the engine that drives local non-farm income opportunities, and when it does not, proximity to urban areas is important (Dorosh and Thurlow 2014). Conversely, many remote farmers in SSA have no access to non-farm income at all (Frelat et al. 2016). In addition, there is evidence that more affluent farmers also have better non-farm opportunities, which weakens the links between non-farm income gains and poverty reduction (Bezu et al. 2012; Haggblade et al. 2010; Djurfeldt and Djurfeldt 2013).

In contrast, technological transformations in agriculture can occur in a single generation. During Asia's Green Revolution, new seeds and new farming practices spread quickly, especially among rice and wheat farmers (David and Otsuka 1994; Evenson and Gollin 2003a). As a result, rural incomes grew directly from non-farm productivity gains. Businesses catering to agriculture and farming households also benefited, spurring growth in non-farm employment. Rural families were able to invest in the health and education of their children, helping them to prepare for jobs in other sectors. In short, Asia's Green Revolution transformed rural economies and engendered a type of economic growth that benefited the poor (Rosegrant and Hazell 2000; Hayami and Kikuchi 2000; Hazell 2009).

Furthermore, the dynamics of Asia's success are globally relevant. A wide range of country and cross-country studies suggest that productivity gains in agriculture are a powerful catalyst for poverty reduction and economic growth (de Janvry and Sadoulet 2010; Irz et al. 2001; Diao et al. 2010; Bravo-Ortega and Lederman 2009; Christiaensen et al. 2011; Anríquez and López 2007; Anderson et al. 2010). Conversely, past efforts to promote other sectors at the expense of agriculture slowed growth and lowered incomes instead (Mundlak et al. 1989; Coeymans and Mundlak

[1] Fox et al. (2013) estimate that agriculture will account for 37% of new jobs in SSA between 2010 and 2020; household enterprises will generate 38% of new jobs, while the formal service and industrial sectors will account for 21% and 4%, respectively.

1992; Bautista and Valdés 1993). It is worth pointing out that the results are consistent across a wide range of farm structures, including the small farms of Africa and Asia and the larger farms of Latin America.

In most places, policies that distort domestic agricultural prices to favour other sectors have waned; however, this is less true in SSA than in other developing regions (Anderson 2009). Using panel data, Anderson and Brückner (2012) show that a continuation of anti-agricultural policy bias continues to slow overall economic growth in the region.[2]

3 Scale, Technology Adoption and Global Food Supplies: Past Lessons and Future Prospects

Despite the many changes brought about by Asia's Green Revolution, sector productivity in Asia is still driven by what happens on small farms, and the same is true in SSA. In East Asia, South Asia and SSA, 95% of the farms are less than 5 ha in size and these farms occupy most of the farmland in these regions (Lowder et al. 2014). Additionally, historical farm census data suggest that the small scale of farming in Asia and Africa persists, even when economic growth in non-agricultural sectors is high. In fact, if there is a noticeable trend, the trend is towards smaller farms (Table 1).

Still, the small scale of farms in Africa need not stand in the way of technology adoption and productivity gains. Indeed, the breakthroughs that launched Green Revolutions in Asia and Latin America largely centred on seeds, not machines, so the benefits were available to farms of all sizes. Nevertheless, initial adoption rates were highest on Asia's small farms, in part because the technologies worked especially well in places where labour was abundant (Hossain 1977). For example, Evenson and Gollin (2003b, p. 450) reports that by 1998, about 82% of the area in Asia planted to major crops used improved seeds. In Latin America, where farms are larger, adoption rates were similar for wheat, a significant export crop; however, rates were lower overall, with 62% of the land planted to modern varieties by 1998.

Furthermore, there is evidence, mostly from Asia, that an agrarian structure composed mainly of small farms is a better foundation for technology diffusion and overall economic growth (Lipton 2009, Chap. 2). For example, Singh (1985) shows that Indian villages with smaller farms and a more equitable distribution of land adopted Green Revolution (staples) and White Revolution (milk) technologies more quickly than otherwise similar villages. Bardhan and Mookherjee (2006) find similar results in West Bengal during the 1980s and 1990s. At a national level, Jeon and Kim (2000) report production and income gains from Korean land reforms carried out in the 1950s that reduced average farm holding size. Using a cross-country panel, Vollrath (2007) finds that output per hectare improves as land distribution becomes more equitable.

---

[2]Conversely, the authors found no evidence that distorting prices to favour agriculture speeds growth.

Table 1 Average farm size from census data (ha)

	Decade				
	1960	1970	1980	1990	2000
India	2.70	2.30	2.00	1.60	1.30
Indonesia	1.20	1.10	1.10	0.90	0.80
Bangladesh	1.40		1.30		0.30
China			0.56	0.43	0.40
Cote d'Ivoire		5.00			3.90
Ethiopia		7.30	4.20	6.20	
Kenya	11.70	4.10	2.50		
Philippines	3.60	3.60	2.90	2.20	2.00
Pakistan	3.50	5.30	4.70	3.80	3.10
Senegal	3.60				4.30
Tanzania		1.30		2.80	2.40
Thailand	3.47		3.70	3.60	
Uganda	3.30			2.20	
Vietnam				0.50	0.70

Source Fan and Chan-Kang (2005), Lowder et al. (2014)

Still, in Africa, the spread of the technologies that launched Asia's Green Revolution stalled. By 1998, only 27% of farmland in SSA was planted to modern varieties. Adoption rates subsequently improved but remained well below rates on Asia's small farms. By 2005, the adoption rates for new varieties were 45% for maize, 26% for rice and 15% for sorghum (Binswanger-Mkhize and McCalla 2010; Pingali 2012).

Nevertheless, gains outside of Africa were sufficient to drive global food markets. During the first Green Revolution, productivity gains from improvements in crop germplasm boosted global agricultural productivity by 1% per year for wheat, 0.8% for rice and 0.7% for maize (Evenson and Gollin 2003a; Pingali 2012). From 1961 to 2001, world maize, rice and wheat yields grew annually at 2.1%, 1.9% and 2.3%, respectively, well above the 1.8% growth in population (FAOSTAT 2015). In Asia, rice yields grew by 2% annually, and maize and wheat yields grew by more than 3%. Productivity gains outpaced demand growth and real prices for food fell.

Since 2000, the global experience with food prices has changed. Real cereal prices, which declined at an annualised rate of 2.3% from 1961 to 2000, rose on average by 6.8% per year between 2000 and 2013; real food prices rose by 5.5% (World Bank Pink Sheet 2015). Additionally, the period was punctuated with sharp price spikes with harsh consequences for the poor. Although food prices subsequently fell, current projected prices remain above the 1990s levels, but below prices at the start of Asia's Green Revolution (World Bank 2015).

Along with prices, concerns are rising that food security gains will be hard to maintain going forward. The global population is expected to grow to 9 billion before levelling off in 2050, and twice as many people are expected to live in Africa in 2050 as in 2010 (Godfray et al. 2010b). Income gains, especially in fast-growing Asia, are driving a still-incomplete nutritional transition that features a growing demand for animal protein and feed crops, which compete with food crops for land (Popkin 1998; Delgado et al. 2008). Furthermore, although they represent a poor choice of instruments to address climate change, biofuel mandates and subsidies have become an integral part of agricultural and energy policies, increasing pressure on agricultural resources (de Gorter and Just 2010).[3]

Using projections for food and biofuel demand by Bruinsma (2009) and Fischer (2009), Hall and Richards (2013) calculate that grain yields on current agricultural lands would need to grow by 1.16–1.31% per year (depending on uncertain policy-driven biofuel demand), to keep pace with projected demand by 2050. The researchers examine known-but-unexploited opportunities to improve yield potentials for rice, maize and wheat, and speculate about promising new technologies, but conclude that current and expected future rates of improvement in genetic potential will be insufficient alone to meet future demand. In general, researchers range from cautious to pessimistic about the ability of new technologies to push the production potential of new seeds to a level sufficient to meet future demand.[4]

Currently, there are signs that productivity growth has stagnated in many areas, falling behind the pace needed to meet future demand, and leaving fewer opportunities for future global gains (Ray et al. 2012, 2013). Poor management of water, fertiliser and pesticides globally, and specifically during Asia's Green Revolution, has degraded land and water resources and has likely played a role in declining yields (Pinstrup-Andersen and Hazell 1985; Pingali and Rosegrant 1994; Godfray et al. 2010a; Stevenson et al. 2013). Moreover, especially in Africa, inadequate fertiliser use has degraded the nutrient content of soils (Cobo et al. 2010; Drechsel et al. 2001). Further ahead are the uncertain consequences of climate change on food production and prices (IPCC 2014). In addition, the anticipated transition to mechanisation in fast-growing countries in Asia, where wages are increasing, may have the unexpected effect of under-cutting future land productivity gains (Otsuka et al. 2016).

Against this background, we discuss two alternative paths to global food security especially relevant for SSA. The first is to expand production by bringing more land and water into agriculture. In general, the potential for bringing more natural resources into agriculture in a sustainable way is limited, especially once agricultural lands lost to urbanisation and competing resource demands are taken into account. Nevertheless, many researchers conclude that there is scope, especially in SSA. An alternative approach is to improve the productivity of existing farms.[5] There are a number of places worldwide where there is significant potential to 'close yield

---

[3]Fischer (2009) estimates that biofuels will add an additional 9–19% to global cereal demand by 2050.

[4]See references in Hall and Richards (2013). See Khush (2013) for an assessment of opportunities for higher yielding rice varieties and Hawkesford et al. (2013) for wheat. Thornton (2010) discusses livestock technologies and the potential for feeding efficiencies.

[5]Godfray et al. (2010a) suggest three additional paths: reducing waste, changing diets and expanding aquaculture.

gaps'—that is, to reduce the gap between productivity obtained and determined potential productivity, by using the best available genetic materials and technologies. However, some of the largest yield gaps are to be found in SSA, where most agricultural resources are managed by smallholder farmers.

4 Smallholder Resources in SSA as a Solution to Future Food Needs

Studies show that staple crop yields could be improved by using better seeds and better nutrients, and by improving water management (Neumann et al. 2010; McDermott et al. 2010; Mueller et al. 2012). This is especially true for SSA, where yield gaps and poverty are linked (Dzanku et al. 2015). For example, in their global study, Mueller et al. (2012) note that some of the largest yield gaps—the difference between potential and actual yields—are for African staple cereals.

Alternatively, it is also possible to expand food production in Africa by planting more land to crops in places where agricultural land remains relatively abundant. Deininger et al. (2011) estimate that more than 200 million hectares in SSA could be converted to rainfed agriculture, roughly 45% of the total area in the world suitable for expansion. Much of the land is isolated; however, they calculate that about 95 million hectares could be accessed without a major investment in infrastructure. In the case of rice, Balasubramanian et al. (2007) estimate that, once double cropping is taken into account, upwards of an additional 236 million hectares of agroclimatically suitable wetlands are available in Africa. Additionally, they argue that the expansion of rice production need not compete with other food crops, since much of the low-lying wetlands suitable for rice are inhospitable for other crops.[6]

Still, converting new lands to agriculture seems like an unlikely path to expanding crop production. On average, the share of agricultural land cropped in Africa stands at 22%, leaving large tracts of meadow and pastureland available to convert to crops with fewer upfront costs to farmers.[7] In addition, there are also important advantages to spending public resources on improving the productivity of existing farms, rather than investing in infrastructure or incentives to carve out new farms. For example, Deininger and Byerlee (2012) estimate that quadrupling maize production through area expansion alone would require 90 million new hectares. Alternatively, decreasing average maize yield gaps in SSA from 80 to 20% would bring about an equivalent increase on existing cropped land, leaving land and water sources to sustain natural

[6] Authors providing estimates of available land are keen to note that the exercise is speculative. One key obstacle is reaching an informed assessment of soil quality (See de Paul Obade and Lal 2013).

[7] In contrast, 87% of agricultural land in Southeast Asia is cropped.

ecosystems and the services they provide, including carbon sequestration (Satterthwaite et al. 2010; Godfray et al. 2010a, b).[8] Furthermore, closing the yield gap by improving soil nutrient management could halt and possibly reverse the problem of declining soil fertility in Africa (Deugd et al. 1998; Place et al. 2003; Zerfu and Larson 2010).

Even so, while there is widespread agreement that boosting productivity on current lands is the preferred approach to expanding food production, there is less agreement on how to close yield gaps, and even disagreement on the relevance of calculated yield gaps. To begin with, potential yields are usually based on selected genetic material combined with inputs, especially chemical fertiliser that must be purchased (Lobell et al. 2009). Tautologically, fully closing the gap means inducing farmers to take up the technologies implicit in calculating the gap. As discussed below, seed-fertiliser technologies work well for some crops in some parts of SSA. In other places, the technologies are simply not adopted. This has led some to question the relevance of conventional yield gap measures and propose alternative measures. For example, Tittonell and Giller (2013) suggest calculating productivity potential based on ecologically intensive farming methods. These methods, which depend on ecological processes to minimise perpetual purchased inputs, can be indicative where fertiliser and other inputs are expensive relative to output value. Tittonell and Giller (2013) note that common elements of ecological intensification, such as integrated cropping and livestock practices, are already present in many African farming systems; however, they also warn that ecologically intensive systems will not suit the needs of all farmers. Marenya and Barret (2007) make the same point, arguing that poor smallholders in western Kenya are unable to afford integrated soil fertility management techniques.

A similar argument applies to water management technologies. Discussions about yield potentials often distinguish between plentiful and scarce water conditions (e.g. Hall and Richards 2013). While it is easy to think about large state-financed irrigation systems that dramatically change water scarcity, there are less capital-intensive water management technologies, such as bunding and levelling, which significantly add to the marginal value of complementary purchased inputs and to productivity overall. As Lipton (2012) points out, this was also the case at the start of Asia's Green Revolution, when farmers often employed simple labour-intensive irrigation systems, including systems that relied on hand-operated pumps and animal-lifted water.

More generally, calculated yield gaps, regardless of methodology, ultimately depend on a technology choice that may or may not be relevant for place-specific livelihood strategies. For this reason, yield gaps are useful as an indicator of prevailing technology choices, relative to a practical standard, but provide less insight into how yield gaps come about or how they might be closed. In Sect. 7, we return to the topic and discuss practical aspects of technology choices and the role of policy.

---

[8]Converting pastureland, though less impactful than carving out new farms, can also have negative environmental consequences, especially when soils or climate make the areas marginal for annual crops. The effects of conversion are place-specific and highly variable. See, for example, Alem and Pavlis (2014), Searchinger et al. (2015).

5 Alternative Rural Development Strategies Based on Larger Farms

The potential for agriculture to grow through expansion, together with dissatisfaction over the uneven pace of yield gains among African smallholders, has given rise to an active debate about smallholder-focused rural development strategies (Lipton 2006; Hazell et al. 2010; Wiggins et al. 2010; Collier and Dercon 2014). Critics argue that small farms forego economies of scale in terms of skills, technology, finance and capital, and face diseconomies in terms of trading, marketing and storage, disadvantages that will become more pronounced as a larger share of the population moves to cities and as rural labour wages converge towards the higher levels found in other sectors (Maxwell and Slater 2003; Collier and Dercon 2014). Consequently, they argue, policies should rely on larger commercial farms to generate productivity gains, while relying on accelerating growth in other sectors to reduce poverty among the rural poor.

The notion that agricultural policies should not neglect larger farms is not controversial. In particular, government efforts to protect security of tenure and property rights and to promote land markets (for rentals and sales) are expected to both benefit smallholders currently and set the stage for an emerging class of medium-sized, commercial farms (Jayne et al. 2014). However, it seems unlikely that shifting budgetary resources away from smallholder farms to support large-farm programmes will help this generation of rural poor or boost global food supplies any time soon.

Farm census, data show there are relatively few large farms in SSA, and there is little to suggest that market forces are driving widespread consolidation (Masters et al. 2013). In addition, there are good conceptual reasons to suggest a smallholder structure is well suited for most places in SSA, where land, labour, risk and credit markets are imperfect (Binswanger and Rosenzweig 1981; Feder 1985).

There are several reasons why Africa's farm structure is based on small farms, but most development economists see two aspects of labour markets as fundamental.[9] The first has to do with supervision and incentives. To start, the incentives to farm diligently and to manage soil and other natural resources are greater for family members than for hired labour. In addition, reliable hired labour can be costly to find and to supervise, which lowers the value of hired labour relative to the value of family labour (Yotopoulos and Lau 1973; Kumar 1979). For this reason, most farms in the world are family-operated—about 75% of all farms according to Lowder et al. (2016). In itself, family ownership is not tied to scale; however, when the transition of labour markets is in its early stages, labour costs are lower in agriculture than in other sectors, which tilts profitable technology choices away from mechanisation and towards labour-intensive technologies. Poor access to credit reinforces this tilt, since farmers can improve their farms through 'sweat equity', for example, by levelling fields or building fences, with little or no credit. Taken together, this means that the best-suited technologies are associated with small farms managed and worked

[9] See Lipton (2009, Chap. 2) for an integrated treatment and an excellent review of the empirical literature.

primarily by household members. Therefore, the most productive farms will be small in most places in SSA. Poorly functioning land markets also work to help keep farms small, because they keep the risks and costs of renting or purchasing land unnecessarily high (Lipton 2009; Deninger et al. 2014).

In many places in Africa, input and output markets are plagued by high transport and transaction costs, which tend to raise farm-gate input prices and lower output prices. In general, this works against high-yielding technologies that rely on purchased inputs, especially fertiliser. It also encourages the adoption of livelihood strategies built around producing some or all of the household's food, since doing so avoids the high transport and transaction costs embedded in purchased food, and discourages the adoption of alternative strategies based on commercial sales, which rely more on markets. This strategic choice comes at a cost, since families may forego planting alternative income-producing crops and forego alternative non-farm activities that are more profitable on average; however, the strategy can prove crucial when food prices surge.[10] Nevertheless, once the decision to produce food for home consumption is taken, price and output risk can lead families on smaller farms to use more household labour and obtain higher yields than neighbours on larger farms (Srinivasan 1972; Barrett 1996).

Regardless of its conceptual underpinnings, there is considerable evidence that productivity, especially land productivity, is higher on small farms than on larger ones in developing countries. Moreover, evidence of an inverse relationship between productivity and farm size is found across a large number of farming systems of varying average scale and under a variety of agroclimatic conditions. [See the literature reviewed in Binswanger et al. (1993), Lipton (2009), and Eastwood et al. (2010)]. In SSA, inverse relationships have been found in Ethiopia (Cornia 1985; Nega et al. 2003), Kenya (Larson et al. 2014), Madagascar (Barrett 1996), Nigeria (Cornia 1985), Malawi (Larson et al. 2014), Rwanda (Byiringiro and Reardon 1996; Ali and Deininger 2015), South Africa (van Zyl et al. 1995), Tanzania (Cornia 1985; Larson et al. 2014), Uganda (Cornia 1985; Nkonya et al. 2004; Matsumoto and Yamano 2013; Larson et al. 2014) and Zambia (Kimhi 2006).[11]

Still, some researchers are sceptical of these results and speculate that they are statistical artefacts. One line of reasoning is that smallholder lands are more productive because farmers are less likely to sell or rent out their highest quality land. As a consequence, productivity differences may be falsely attributed to scale, since land quality usually goes unmeasured (Assunção and Braido 2007; Benjamin 1995; Bhalla and Roy 1988; Lamb 2003). Potentially, systematic measurement errors in

---

[10] See Yamauchi and Larson (2016) and references therein for a discussion about the effects of food price crises on child health.

[11] An inverse relationship is often found outside of SSA, where the average scale of farms is more varied. For example, inverse relationships have been found in Barbados (Cornia 1985), Brazil (Kutcher and Scandizzo 1981; Berry 1984), the Indian Punjab (Sen 1966), Malaysia (Berry 1984), Mexico (Cornia 1985), the Philippines (Hayami et al. 1990), the Republic of Korea (Cornia 1985) and West Bengal (Carter 1984).

self-reported area and yields may also bias empirical results (Lamb 2003). The empirical evidence addressing either criticism is thin, but a study by Barrett et al. (2010) that includes soil measurements finds no evidence of systematic bias for measured yields. Matsumoto and Yamano (2013) found inverse relationships between maize yields and plot size from estimated models that include and exclude soil carbon levels. Additionally, a recent study that incorporates self-reported and GPS-measured plot areas in Uganda concludes that measurement errors work against, rather than in favour of, the inverse yield hypothesis (Carletto et al. 2013).

Some critics discount the inverse productivity findings, arguing that the empirical samples behind the studies are composed mainly of small farms (Collier and Dercon 2014). This latter criticism is unfounded, especially when studies outside of Africa are considered. More importantly, the criticism also misses the larger point that, among the applied technologies used by a large representative class of farmers, there is no evidence that significant production economies are given up by staying small. In fact, the empirical evidence suggests the opposite: that the small-scale structure of SSA agriculture is an efficient way to use land and labour resources, given prevailing market conditions and constraints. For policy, this means that the full benefit of scale-neutral technology innovations, such as the development of improved seeds, can be achieved without augmenting the limited set of assets, mostly land and family labour, that smallholder farmers already possess. This is not to say that African farms should stay small, but it does offer a path for reducing poverty and building global food supplies without waiting for labour markets, capital markets and farm structures to change.

As an alternative to smallholder-centred policies, some researchers suggest promoting larger commercial farms on new land, by taking advantage of investor interest in acquiring large tracts of agricultural lands (Collier and Dercon 2014). The strategy is risky, and even proponents urge taking a cautious and experimental approach (Collier and Venables 2012). This is because the history of government-managed land transfers is poor, especially in Africa (Eicher and Baker 1982; Andrae and Beckman 1985; Zoomers 2010; Borras et al. 2011; Deininger and Byerlee 2012). Moreover, Arezki et al. (2015) report worrisome evidence that current interest in African land deals is highest where land governance and tenure security are weak.[12]

Given the need for caution, programmes to develop new farms are unlikely to have a significant impact on food supplies any time soon and seem a less strategic use of resources when compared to programmes designed to achieve small increases in the yields of current farmers. Furthermore, as discussed, improving smallholder yields would likely generate greater economic growth and larger reductions in poverty than large-farm strategies, even when the effects on global food supplies are equivalent.

[12]Byerlee et al. (2017) document an alternative approach where developers invest in improving frontier lands with the intention of selling subdivided parcels to individual family-run farms. The approach resolves upfront investment hurdles while ultimately resulting in the creation of new family-scaled farms.

6 Revolution or Evolution?

Many of the comparisons between Africa now and tropical Asia prior to its Green Revolution are apt. At the start of Asia's Green Revolution, agriculture was structured around small farms. Yields of the two main staple crops, rice and wheat, had been stagnant while populations grew. Consequently, food insecurity and fear of famine were widespread (Meadows et al. 1972; Drèze and Sen 1989; Otsuka and Place 2015). However, in Asia, new technologies—once proven—were quickly taken up with exceptional results. In contrast, progress in Africa has been characterised by a mosaic of local successes that are hard to detect in national or regional numbers (Reij and Smaling 2008; Otsuka and Larson 2013).[13] In turn, this has fed disappointment among donors with policies meant to improve smallholder productivity and has promoted a view that the smallholder policy narrative has been oversold (Dercon 2013).

However, there are also key differences that make an African Green Revolution harder to achieve. First is the diversity of agroclimatic-based food systems and the related diversity in the staple crops that are the foundation of African diets. This point is illustrated in the top panel of Fig. 1, which shows the share of calories originating in the two major grains for SSA (rice and maize) in 2010 and also the share of calories originating in wheat and rice for South and Southeast Asia, at the start of Asia's Green Revolution.[14] The country values presented in the figures are ranked from lowest to highest. In Asia, food systems had emerged based on large east–west-oriented agroclimatic zones, well suited for wheat and rice. Additionally, weather variations were mitigated by investments in irrigation.[15] By 1965, more than 46% of the calories available in every country of South and Southeast Asia came from rice or wheat; in populous Thailand, Vietnam and Bangladesh, the shares exceeded 65%. Consequently, technological innovations in wheat and rice, once adopted, had the potential for large system-wide impacts on productivity, incomes and nutrition.

In contrast, the food systems in Africa, supported by largely north-south oriented agroclimatic zones, are more varied. In 2010, only food systems in Malawi, Zambia, Lesotho and Madagascar depended mostly on maize and rice. Consequently, the successful adoption of a single high-yielding crop technology—for example, high-yielding lowland rice—can be important regionally without adding up to system-wide impacts. In other words, the number of crop innovations must be greater in Africa to generate a continental Green Revolution.

Another significant difference has to do with land availability. The lower panel of Fig. 1 shows the portion of agricultural land planted to annual or permanent crops,

---

[13]Lipton (2012) makes a similar point citing more accurately measured weight-and-height data.

[14]Cassava is another important crop and one that, historically, has been more important than rice in SSA. Statistics on cassava consumption are prone to error, since a large share of domestic production is not marketed. However, FAOSTAT (2015) data for 2010 suggest that rice generates about 244 calories per capita per day for the region, compared to 191 calories for cassava.

[15]In 1965, about 12% of cropland in Southeast Asia and 20% in South Asia was equipped for irrigation. In contrast, only 3% of cropland in SSA was equipped for irrigation in 2013.

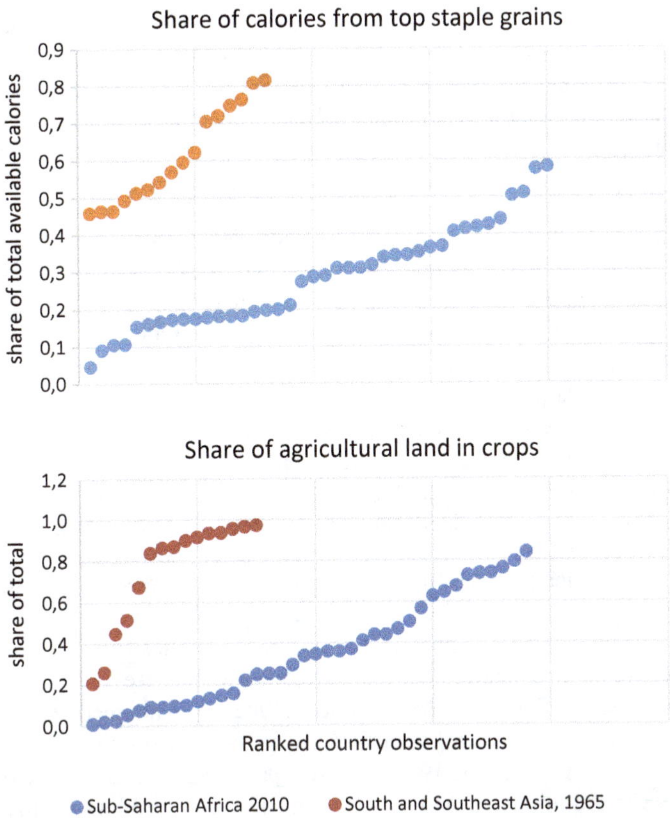

Fig. 1 Starting points for South and Southeast Asia, and Sub-Saharan Africa. *Source* FAOSTAT (2015); authors' calculations. *Notes* The figures plot ranked country average values from lowest to highest. For SSA, reported calories originate in rice, maize. For South and Southeast Asia, the reported calories originate in rice, wheat

with the remaining portion devoted to meadows and pastureland. In the places where Asia's Green Revolution found early success, such as Pakistan, India and Thailand, more than 85% of available agricultural land was already cropped, leaving little room to expand by converting pastureland. In Asia, the technologies that proved successful were fertiliser-responsive high-yielding varieties of rice and wheat, which greatly improved land productivity. Because household landholdings were also small, the seed-fertiliser technologies solved a constraint faced by many farming households, as well as a primary constraint for the sector as a whole (Johnston and Cownie 1969).

Land availability is more varied in SSA. In places such as Mozambique and Liberia, large tracts of pastureland are available for conversion. But more than 74% of the land is already cropped in Burundi, Rwanda and Cameroon. Consequently, technologies that boost land productivity are more relevant in some places than others. In addition, differences in transportation and transaction costs are often greater in more sparsely populated land-abundant countries. This also affects technology choices, since remote households farming where input prices are higher and output prices lower will find fertiliser-intensive technologies less attractive (Larson and Gurara 2013). This point is brought home by considerable empirical evidence that farmers' preferences for varieties reflect more than potential yields (Adesina and Baidu-Forson 1995; de Groote et al. 2002; Lunduka et al. 2012).[16]

The diversity of economic and agroclimatic conditions, and its implications for the suitability of any single technology, also complicates the design and implementation of programmes meant to disseminate developed technologies. In Asia, national campaigns to improve smallholder productivity could be built around one or two crops, with standard knowledge bases. Moreover, lessons gained from the same crops in one country could be adopted in the next. In Africa, regional campaigns with multiple technologies and varying knowledge bases are needed to achieve the same aggregate impact. In addition, research and extension successes in a handful of populous countries in Asia led to significant global impacts, while success in SSA depends on making headway across 49 independent national research and extension systems.

In addition, Africa's Green Revolution got off to a slow start. The foundation for Asia's early success was already in place by 1965, based on a large stock of improved germplasm from temperate zones, wheat varieties from North America, Europe and Japan, and rice cultivars derived from Taiwanese and Japanese semi-dwarfs (Evenson and Gollin 2003c). In contrast, international breeding programmes for sorghum, millet, barley, lentils, potatoes and cassava—crops that are important in Africa—did not begin until the 1970s, and rice programmes for Africa did not start until the 1980s. Additionally, African governments have been less willing to invest in agriculture in general, and in agricultural research specifically. Based on available data, Pardey et al. (2006) note that 44 countries in SSA spent a combined USD 3.8 billion on agricultural research in 2000. By way of contrast, India invested almost USD 21 billion and China invested more than USD 48 billion during the same year.[17]

Because of the great diversity of circumstances that condition smallholder technology choices, the smaller initial portfolio of relevant technologies, and a history of meagre support for agricultural research, Africa's Green Revolution has been slower to evolve than Asia's. It has also meant that successes to date have been dispersed and

---

[16]For example, Diagne et al. (2013) posit that some farmers were attracted to Nerica rice because of its shorter growing season despite similar observed yields. In Burkina Faso, Dao et al. (2015) find seed colour and drought resistance are important factors for maize farmers. The researchers also find regional differences in preferred traits.

[17]In 2003, African governments entered into a pledge, known as the Maputo Declaration, to allocate 10% of government expenditures to agriculture in order to address past neglect. Benin and Yu (2013) report that progress towards meeting pledged goals has been mixed.

[18]National governments and the international community have also under-invested in gathering agricultural statistics, which makes international comparisons indicative at best. Furthermore, better measured yields from households are not representative of national outcomes, since they are drawn from a sample of households rather than farms.

the cumulative effects of success harder to discern. Even so, there are signs that the gap in cereal yields is beginning to close in some places.[18] Figure 2 shows annualised

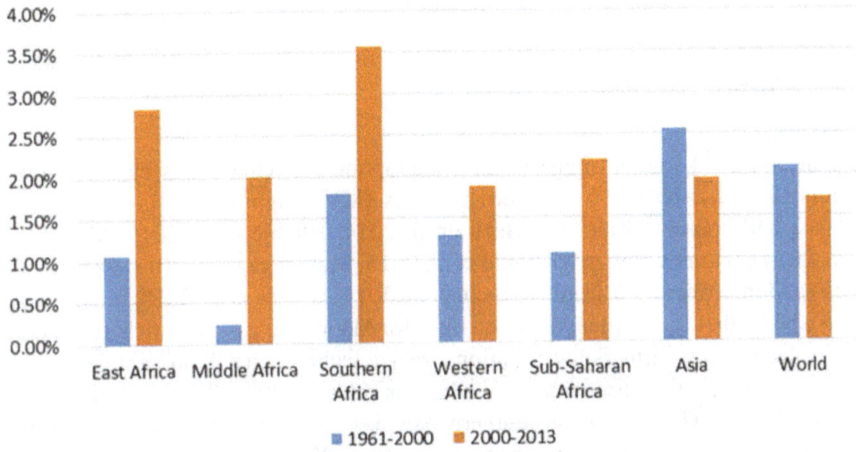

Fig. 2 Annualised rates of yield growth for cereals, Africa, Asia, and World. *Source* FAOSTAT (2015); authors' calculations. *Note* East Africa includes: Burundi, Comoros, Djibouti, Kenya, Madagascar, Malawi, Mauritius, Mozambique, Eritrea, Zimbabwe, Réunion, Rwanda, Somalia, United Republic of Tanzania, Uganda, Ethiopia and Zambia. Middle Africa includes: Angola, Cameroon, Central African Republic, Chad, Congo, Gabon, Sao Tome, Principe, and the Democratic Republic of the Congo. Southern Africa includes: Botswana, Lesotho, Namibia, South Africa and Swaziland. West Africa includes: Cabo Verde, Benin, Gambia, Ghana, Guinea, Côte d'Ivoire, Liberia, Mali, Mauritania, Niger, Nigeria, Guinea-Bissau, Senegal, Sierra Leone, Togo and Burkina Faso

growth rates in yields for two periods: the twentieth century Green Revolution years 1961–2000 and the more recent period 2000–2013. During the first period, growth in Asian cereal yield outpaced that in SSA by a wide margin. Since then, Asian yield growth rates have slowed somewhat, as have average world rates. However, growth rates in SSA have accelerated. The figure shows considerable regional differences for both periods, with the largest yield gains occurring in Southern Africa. Keeping in mind that yield improvements in Asia have been sustained for six decades, it is clear that the yield gap between Asia and SSA remains large. However, the recent differential growth rates in crop yields do suggest that the gap between Asia and parts of Africa have begun to close.

7 Lessons from Farmers' Fields

In this section, we focus on elements of maize and rice productivity in SSA that illustrate why the task of developing and disseminating appropriate smallholder technologies is complex in SSA, and provide some examples of technologies that have succeeded. The section draws on seven country studies focused on irrigated and non-irrigated rice technologies, rainfed maize technologies, irrigation systems management, training and credit programmes. The detailed studies are reported in Otsuka and Larson (2016).

7.1 Maize

Recent empirical studies suggest that the adoption of technologies built on high-yielding fertiliser-responsive modern varieties (MVs) significantly increases crop revenue and household income (Bezu et al. 2014; Khonje et al. 2015; Mathenge et al. 2014), reduces poverty (Khonje et al. 2015; Mathenge et al. 2014; Zeng et al. 2015) and improves food security (Bezu et al. 2014; Khonje et al. 2015; Smale et al. 2015). Nevertheless, regional adoption rates for MV maize vary significantly, and in many places, most farmers use traditional maize technologies. For example, Smale et al. (2013) report the following MV adoption rates for 2006–2007: 33% of maize area in Eastern Africa; 52% in Southern Africa, excluding South Africa; and 60% in West and Central Africa. Drawing on data from 2001 to 2006, Kostandini et al. (2013) also report considerable variations in MV adoption rates among countries, ranging from 5% in Angola to over 70% in Kenya and Zambia.[19]

As in Asia, the full benefits of the maize technologies developed for Africa are only achieved with high levels of soil nutrients, which in practice calls for the use of chemical fertilisers (Morris et al. 2007). There is ample evidence that the MV technologies, once adopted and used in combination with fertiliser, perform as expected. For example, estimated marginal response rates of maize to fertiliser range from 12 kg maize per kg of nitrogen application in Zambia (Xu et al. 2009) to 25 kg in Kenya (Matsumoto and Yamano 2013). In addition, studies suggest that adopting seed-fertiliser maize technologies is profitable in: Western, Eastern and Southern Africa (Morris et al. 2007); Kenya (Sheahan et al. 2013; Matsumoto and Yamano (2013); Uganda (Matsumoto and Yamano 2013); and Zambia (Xu et al. 2009). Yet the share of farmers that apply fertiliser to maize plots varies widely. For example, Sheahan et al. (2013) report use rates of 76% in Kenya in 2010; Matsumoto and Yamano (2013) report use rates of 74% in central and western Kenya in 2004 and 2007. In neighbouring Uganda, Larson et al. (2016) found that 6% of maize farmers used fertiliser in 2009–2010; Matsumoto and Yamano (2013) report a 3% share.

As discussed, boosting the adoption of MV technologies is often the key proximate objective of rural development strategies, so the question of why adoption rates are often low among African maize farmers has received considerable attention. Recently, discussions have focused on rates of fertiliser use, which are low when compared to recommended levels and in comparison to typical practices on Asia's

[19]Regional and national MV adoption rates were extracted from DTMA seed sector surveys for Eastern African and Southern Africa (Langyintuo et al. 2010) and from Alene et al. (2009)'s own survey. Both data sets are based on interview surveys from maize seed companies, national research organisations and community-based organisations.

small farms. In the case of maize, there is often evidence that MV technologies are welfare improving, so much of the recent debate has focused not on the technology promoted, but on information gaps, market failings and 'non-fully rational behaviour' (Larson et al. 2016; Duflo et al. 2008). Generally, policy advocates agree on the need for governments to invest more in infrastructure and other public goods, which tend to lower barriers to technology and markets, but there is less agreement about whether additional interventions are needed.

7.1.1 Appropriate Technology but Inadequate Promotional Programmes?

One area of debate is whether the market reforms eliminating grain and fertiliser parastatals, which accelerated in the 1990s, stunted the adoption of high-yielding varieties (Akiyama et al. 2003; Crawford et al. 2003; Jayne et al. 2003; Morris et al. 2007). A related debate concerns the importance of supported output prices and subsidies on fertiliser, irrigation, power and credit during the early stages of Asia's Green Revolution (Timmer 1997; Johnson et al. 2003). Some conclude that markets have failed in Africa and call for greater state intervention (Moseley et al. 2010; Poulton et al. 2006; Winter-Nelson and Temu 2005). Still, in the case of maize, there have been several aggressive state-sponsored campaigns reminiscent of the early programmes in Asia, and, more recently, national programmes built around subsidised inputs (Jayne and Rashid 2013).

Perhaps the effort most comparable to the Asian programmes was the Global 2000 Campaign, promoted by Norman Borlaug (who had been awarded a Nobel Prize for his work on Asia's Green Revolution), former US President Jimmy Carter, and Japanese billionaire philanthropist Ryoichi Sasakawa. The regional programme, begun in Ghana in 1986, was designed to promote the hands-on transfer of MV maize technology by offering a selected package of seeds, fertiliser and extension services to a small cadre of farmers (Brinkley 1996). In Ghana, maize yields tripled on early demonstration plots, and by 1989, nearly 80,000 farmers had adopted the recommended MV technology. Nevertheless, Tripp (1993) argues that many of the technologies promoted were not consistently the best technologies in all regions of Ghana and were often marginally profitable under ideal conditions. When poor weather inevitably arrived, many farmers were unable to repay loans, and soon many of the farmers who had adopted the technology abandoned it. A 1990 survey found that only 29% of the farmers who first used fertiliser under the programme continued to use it (Tripp 1993, p. 2012).

Similar programmes were launched for maize in Ethiopia, Ghana, Kenya, Malawi, Mozambique, Sudan, Zambia and Zimbabwe (Brinkley 1996; Howard et al. 2003; Smale and Jayne 2010; Smale 1994). Some programmes showed early success before reaching an adoption plateau; however, adoption gains were often reversed and new technologies abandoned. Frequently, adoption depended heavily on controlled prices or subsidies that proved too expensive to maintain. Similarly, researchers are finding that several recently launched campaigns to promote MV technologies with subsides are proving expensive with at best marginal benefits (Ricker-Gilbert et al. 2013; Jayne and Rashid 2013).[20]

7.1.2 Heterogeneous Conditions, Livelihood Strategies and Technologies

An alternative explanation is that the circumstances that affect the profitability of a given technology are heterogeneous, so it is unlikely that all farmers will decide to use the same technology. In this sense, the choice of technology is endogenous (Mundlak 1988; Mundlak et al. 2012). Differences in natural endowments, for example, soil fertility or water availability, affect fertiliser response rates and yields, and therefore the profitability of a given technology (Larson and Leon 2006; Marenya and Barrett 2009; Jayne and Rashid 2013). Variable weather may add to the risk of a technology, even if the technology is profitable on average (Larson et al. 2016); however, price and weather risks can encourage the use of inputs, especially family labour, which affect the relative productivity of technology choices (Srinivasan 1972; Barrett 1996). Spatial differences in transport and transaction costs also affect relative profitability (Xu et al. 2009). In the specific case of maize, Suri (2011) concludes that in Kenya, differences in the costs and benefits of using maize hybrids explain the differences in choices made by adopters and non-adopters of MV technologies. In short, there are differences in the market prices and relevant shadow prices faced by farmers, which generate variations in livelihood strategies pursued and technologies employed.[21]

Practical differences in technology preferences are illustrated by Table 2, which shows that input intensities such as chemical fertiliser and manure application per hectare, the share of intercropped fields, and the percentage of adoption of hybrid seeds are all substantially higher in Kenya than in Uganda, which resulted in higher maize yields and higher crop income in Kenya than in Uganda. Nevertheless, in the case of Uganda, the use of low input, low yield technologies are consistent with weather-related yield risks (Larson et al. 2016).

7.1.3 The Integrated Maize Farming System

While research and extension programmes often focus on high-yielding seed-fertiliser technologies, another approach is to adopt integrated soil fertility manage-

[20] Jayne and Rashid (2013) report that the 10 African countries covered in their review spent about 29% of their agricultural budgets on input subsidies in 2011.

[21] Moreover, conditions are not static but evolve, especially as urbanisation generates new demand and new markets for food. See, for example, the discussion about the role of markets in Haggblade and Hazell (2010).

ment systems, which use both organic and inorganic fertilisers. For example, in India, Kajisa and Palanichamy (2013) found that application of organic fertiliser directly

Table 2 Maize technology adoption in Kenya and Uganda

	Kenya		Uganda
	2004	2012	2009/2010
Maize yield (tons/ha)	1.8	2.1	1.2
Crop income (KSh/ha)	37,869	46,786	–
Chemical fertiliser use (kg/ha)	49	47	0.3
Manure use (kg/ha)	971	1578	22
Share of intercropped fields (%)	76	72	45
Adoption of hybrid maize (%)	50	78	30

Note Adoption rates are based on share of farmers surveyed. *Source* Larson et al. (2016), Muraoka et al. (2016). Crop income is defined as crop production minus all paid costs associated with crop production. The Uganda data is nationally representative of households. Data for Kenya is based on household surveys drawn from the west and central highlands

increased productivity of upland cereals and also raised the efficacy of chemical fertilisers in low fertility soils. Evidence from Malawi and Kenya shows that intercropping or rotating maize with nitrogen-fixing legumes is another possible way to restore soil nutrients (Ojiem et al. 2014; Snapp et al. 2010). In Central and West Africa, farmers fallow land to restore soil fertility and use biological indicators to judge fertility restoration levels (Norgrove and Hauser 2016).

A more complex integrated maize technology, which seems to work well in populous areas that are also free from sleeping sickness, is examined by Muraoka et al. (2016) and Otsuka and Yamano (2005). In the highlands of Kenya, farmers who adopt this system grow forage crops such as Napier grass and feed it to genetically improved stall-fed cows, from which they produce manure. They apply the manure along with chemical fertilisers to fields planted with hybrid maize varieties, often intercropped with nitrogen-fixing legumes. This integrated maize–livestock farming system is more labour-intensive and land-saving than the traditional farming system, in which livestock is grazed without much interaction with crop farming.

It is worth noting that this highly complex farming system is indigenous and evolved as the markets for milk, seeds and fertiliser developed and improved. It is a novel combination of technologies that has not been researched as a system, so the best combination of inputs and the optimum timing of their application is undocumented.

7.2 Rice

This section focuses on a series of empirical studies of rice farming, based on household data collected in Mozambique (Chokwe irrigation scheme in the south and rainfed areas in the central region), Tanzania (three major rice-growing regions), Uganda (lowland rice-growing areas in the east and north), Ghana (rainfed areas in northern Ghana) and Senegal (irrigated areas in the Senegal River Valley), published in Otsuka and Larson (2016). In general, the surveys cover major lowland rice-growing areas in each country. As the studies reveal, lowland rice is a promising crop in Africa and one well suited to technologies currently used in Asia; however, the same is not true of upland rice, which is popular in West Africa. (Otsuka and Larson 2013; Estudillo and Otsuka 2013; Nakano et al. 2013).

In the Senegal River Valley, paddy yields average 4.5 t per hectare, the highest among the five countries studied. Rice farmers in the Senegal River Valley have access to an irrigation facility with ample supply of water, adopt Asian-type semi-dwarf modern varieties, apply chemical fertilisers abundantly, and practise improved management such as bunding, levelling and straight-row planting (Sakurai 2016). Yields are comparable with yields observed in Asia's irrigated areas in the late 1980s (David and Otsuka 1994). Irrigated yields in Tanzania were lower, averaging 3.7 t per hectare, partly because of lower-quality irrigation facilities and the incomplete adoption of improved management practices (Nakano et al. 2016). Irrigated yields in Mozambique ranged between 1.6 and 2.0 t per hectare, well below the average of 2.2 t in SSA (FAOSTAT 2015). This lower yield can be attributed to low-quality irrigation facilities, inadequate application of chemical fertiliser, use of old varieties, and low adoption of improved management practices (Kajisa 2016). Thus, it is clear that irrigation alone is not sufficient to achieve high rice yields.[22]

Njeru et al. (2016) observed paddy yield as high as 5 t per hectare in the Mwea irrigation scheme in Kenya, where improved management practices were widely adopted, even though high-quality but low-yielding basmati varieties are grown. Rice markets are well developed there. Rice is strictly graded, and credits are available from credit unions and also from private traders. In some areas of the Mwea irrigation scheme, where varieties developed by the International Rice Research Institute (IRRI) are grown, yields of 8 t per hectare were obtained. It is thus possible to achieve high yields in irrigated areas in Africa if improved varieties and chemical fertiliser are combined with improved management practices.

Nonetheless, achieving a sweeping revolution in rice will require improvements in rainfed rice yields, since 85% of lowland paddy field in SSA is rainfed (Balasubramanian et al. 2007). Rainfed yields ranged from 0.8 to 1.0 t per hectare in Mozambique, 2.0 t per hectare in Ghana and 2.3–2.5 t per hectare in Uganda. The higher average yields in Uganda are close to the average yield in rainfed areas in Asia in the late 1980s (David and Otsuka 1994). It is important to note that management training programmes were not implemented in Mozambique, but they were in rainfed areas

[22] Yield data in this paragraph are self-reported by farmers.

in Uganda and Ghana.[23] In rainfed areas in northern Ghana, yields averaged a mere 1.5 t per hectare without any improved management practices, but reached 2.6 t per hectare with the adoption of all recommended management practices, including use of improved seeds, chemical fertiliser, bunding, levelling and straight-row dibbling. Similarly, management training is found to be effective in rainfed areas in Uganda. It is noteworthy that substantial yield gains have been achieved, even in rainfed areas in SSA, without accompanying major market reforms or new investments in roads or irrigation systems. Moreover, if the markets work to support the improved rice farming system, further improvement in productivity can likely be achieved.

In Table 3, we turn to a broader measure of productivity and examine the effects of improved management practices on profitability, defined as the value of production per hectare minus paid-out cost and the imputed costs of owned resources evaluated at market prices. Because imputing the value of family labour is prone to error, the table also includes income per hectare, which is defined as the value of production minus paid-out cost per hectare. In Tanzania, both income and profit per hectare were significantly higher in irrigated areas compared to rainfed areas. In rainfed areas in Ghana, both income and profit per hectare were significantly higher for full adopters of improved seeds, fertiliser, bunding, levelling and dibbling than for non-adopters. In Uganda, income per hectare was significantly higher for management training participants. The Ugandan study also found that training participants adopted improved seeds, chemical fertiliser and improved management practices more often than non-participants.[24]

Table 3 Income and profit from rice cultivation, by status of training and technology adoption

	Income per ha	Family labour cost per ha	Profit per ha
Tanzania			
Irrigated area	1011	421	590
Rainfed area	453	300	153
Uganda (rainfed)			
Training participants	1327	–	–
Non-participants	905	–	–
Ghana (rainfed)			
Full adopters	374	215	160
Non-adopters	228	169	59

Sources Kijima (2016), deGraft-Johnson et al. (2016), Nakano et al. (2016), Otsuka and Larson (2016). *Note* income and profit are measured in USD. Income is defined as the value of production minus paid-out costs. Profit is defined as income minus imputed costs of owned resources, including family labour

[23]No subsidy was provided in Uganda, whereas a subsidy for fertiliser and seeds was provided initially in Ghana around 2000.

[24]These results are robust when the treatment effects models were applied (Kijima 2016; deGraft-Johnson et al. 2016).

Considerable differences in income and profit between training participants and non-participants, and between full adopters and non-adopters, suggest that it is profitable to adopt seed-fertiliser-management (SFM) rice technology in SSA. Supporting evidence is provided by an ongoing study into the adoption of intensive rice cultivation systems in rainfed areas in Kilombero Valley in Tanzania by Nakano, Kajisa and Otsuka (2015), who also found significantly higher yields and profit per hectare associated with SFM technologies. Thus, returns on training programme seem very high, although rates of return have not been reported by these studies.

8 Concluding Discussion

Most rural development strategies in SSA focus on improving the productivity of smallholder farms. The approach has clear advantages, since it leverages resources already in place in rural communities—primarily land, family labour, farming knowledge and social capital—while the slow-paced dynamics of economic transformation and farm restructuring play out. Importantly, improving smallholder productivity speeds up economic growth and complements programmes that prepare rural workers for jobs in other sectors.

With the possible exception of mechanisation, technologies that improve overall small-farm productivity usually help farmers to achieve higher yields. In general, yield gaps are high in SSA and even a partial narrowing could have a significant impact on local and global food supplies. A careful and transparent approach to promoting new commercial farms, carved out from under-utilised land resources, can help to improve food supplies over time, but likely with fewer benefits for economic growth or poverty reduction and with greater stress on Africa's natural resources and ecologies.

A key instrument in the pursuit of improving smallholder productivity is the development and dissemination of new technologies. In the case of Asia's Green Revolution, a small set of technologies had a transformational impact on rural communities and food systems, which were land-constrained and structured largely around wheat and rice. In SSA, the agroclimatic and market conditions that support food systems are diverse. Consequently, a larger set of place-specific technologies are needed to prompt an African Green Revolution. This raises the cost of developing and testing new technologies and, importantly, the cost of disseminating technologies that meet farmers' needs.

Lessons from the fields of maize and rice farmers in SSA illustrate these points. In the case of lowland rice, the seed-fertiliser technologies that transformed many Asian food systems work successfully in well-managed irrigation systems and generate similar yields. However, the technologies are not enough to materially change the lives of most rice producers, since most of the rice produced in SSA is rainfed. Nevertheless, research suggests that, in many communities, rainfed rice yields can be improved by focusing not only on seeds and fertiliser, but on improved agronomic practices. It is worth keeping in mind that while the key technologies that drove Asia's

Green Revolution relied on purchased inputs, they also leveraged available family labour. In places where input prices are high, ecologically intensive technologies, such as those based on improved agronomic and resource conservation practices, can further leverage family labour to achieve higher productivity levels with fewer purchased inputs.

In the case of maize, evolving differences in risk, markets and growing conditions help to explain dramatic differences among farmers in terms of use of fertiliser and high-yielding seeds. Heterogeneity also helps to explain why technology promotion programmes based on a single technology packet, which proved transformational in Asia, have seen limited success in Africa. However, as the example of highland maize farmers in Kenya illustrates, indigenous hybrid systems based on improved genetic material and improved resource management can supplement traditional seed-fertiliser technologies.

Taken together, all of this suggests that a strategy of boosting smallholder productivity is a sound one, especially when combined with policies that help families to prepare for jobs outside of agriculture. However, the task of building out, cataloguing and disseminating the full set of technologies needed for transformational change represents a challenge to African governments and the development community. However, there is evidence of local success and some indication that yield gaps have closed in recent years. Pursuing additional paths, built on resource management technologies and indigenous innovations, can speed Africa's Green Evolution.

Acknowledgements This article originally appeared under the title 'Why African rural development strategies must depend on small farms' in *Global Food Security*, Vol. 10 2016, pp. 39–51. The authors thank *Global Food Security*'s editor, Derek Byerlee, and Elsevier for permission to republish the essay in this volume. The study itself is a result of a research project conducted at the Japan International Cooperation Agency (JICA) Research Institute to empirically analyse how best the Coalition for African Rice Development (CARD) initiative can serve to increase rice productivity and reduce poverty in SSA. CARD, which aimed to double rice production from 2008 to 2018, was jointly launched by JICA and the Alliance for the Green Revolution in Africa (AGRA) at the 4th Tokyo International Conference on African Development (TICAD) meeting in 2008. We would like to thank the JICA Research Institute for the intellectual and financial support it has provided for this project. We are also grateful for the financial support provided to maize research in Kenya by the Global Center of Excellence Program, the GRIPS Emerging State Project of the Japan Society for the Promotion of Science (JSPS KAKENHI Grant number 25101002) and the generous support of the donor-funded Knowledge for Change Program hosted by the World Bank.

References

Adesina, A. A., & Baidu-Forson, J. (1995). Farmers' perceptions and adoption of new agricultural technology: Evidence from analysis in Burkina Faso and Guinea, West Africa. *Agricultural Economics, 13*(1), 1–9.

Akiyama, T., Baffes, J., Larson, D. F., & Varangis, P. (2003). Commodity market reform in Africa: Some recent experience. *Economics System, 27*(1), 83–115.

Alem, S., & Pavlis, J. (2014). Conversion of grazing land into *Grevillea robusta* plantation and exclosure: Impacts on soil nutrients and soil organic carbon. *Environmental Monitoring and Assessment, 186*(7), 4331–4341.

Alene, A. D., Menkir, A., Ajala, S. O., Badu-Apraku, B., Olanrewaju, A. S., Manyong, V. M., et al. (2009). The economic and poverty impacts of maize research in West and Central Africa. *Agricultural Economics, 40*(5), 535–550.

Ali, D. A., & Deininger, K. (2015). Is there a farm size–productivity relationship in African agriculture? Evidence from Rwanda. *Land Economics, 91*(2), 317–343.

Anderson, K. (2009). Distortions to agricultural versus nonagricultural producer incentives. *Annual Review of Resource Economics, 1*(1), 55–74.

Anderson, K., Cockburn, J., & Martin, W. (Eds.). (2010). *Agricultural price distortions, inequality, and poverty*. Washington: World Bank.

Anderson, K., & Brückner, M. (2012). *Distortions to agriculture and economic growth in Sub-Saharan Africa* (Policy Research Working Paper 6206). Washington: World Bank.

Andrae, G., & Beckman, B. (1985). *The wheat trap: Bread and underdevelopment in Nigeria*. London and Uppsala: Scandinavian Institute for African Studies, Zed Books Ltd.

Anríquez, G., & López, R. (2007). Agricultural growth and poverty in an archetypical middle income country: Chile 1987–2003. *Agricultural Economics, 36*(2), 191–202.

Arezki, R., Deininger, K., & Selod, H. (2015). What drives the global "land rush"? *The World Bank Economic Review, 29*(2), 207–233.

Assunção, J. J., & Braido, L. H. (2007). Testing household-specific explanations for the inverse productivity relationship. *American Journal of Agricultural Economics, 89*(4), 980–990.

Balasubramanian, V., Sie, M., Hijmans, R. J., & Otsuka, K. (2007). Increasing rice production in Sub-Saharan Africa: Challenges and opportunities. *Advances in Agronomy, 94*, 55–133.

Barrett, C. B. (1996). On price risk and the inverse farm size-productivity relationship. *Journal of Development Economics, 51*(2), 193–215.

Barrett, C. B., Bellemare, M. F., & Hou, J. Y. (2010). Reconsidering conventional explanations of the inverse productivity–size relationship. *World Development, 38*(1), 88–97.

Bautista, R. M., & Valdés, A. (1993). *The bias against agriculture: Trade and macroeconomic policies in developing countries*. San Francisco: International Centre for Economic Growth.

Benin, S., & Yu, B. (2013). *Trends in public agricultural expenditures in Africa*. Washington: International Food Policy Research Institute (ReSAKSS Issue Note 22).

Benjamin, D. (1995). Can unobserved land quality explain the inverse productivity relationship? *Journal of Development Economics, 46*(1), 51–84.

Berry, A. (1984). Land reform and the adequacy of world food production. In J. Montgomery (Ed.), *International dimensions of land reform*. Boulder: Westview Press.

Bezu, S., Barrett, C. B., & Holden, S. T. (2012). Does the nonfarm economy offer pathways for upward mobility? Evidence from a panel data study in Ethiopia. *World Development, 40*(8), 1634–1646.

Bezu, S., Kassie, G. T., & Shiferaw, B. (2014). Impact of improved maize adoption on welfare of farm households in Malawi: A panel data analysis. *World Development, 59*, 120–131.

Bhalla, S. S., & Roy, P. (1988). Mis-specification in farm productivity analysis: The role of land quality. *Oxford Economic Papers, 40*(1), 55–73.

Bardhan, P., & Mookherjee, D. (2006). Land reform, decentralized governance, and rural development in West Bengal. In *Conference on Challenges of Economic Policy Reform in Asia* (Vol. 31).

Binswanger, H., & Rosenzweig, M. (1981). *Contractual arrangements, employment and wages in rural labor markets: A critical review* (Studies in employment and rural development No. 67). Washington: World Bank.

Binswanger, H., Deininger, K., & Feder, M. (1993). *Power, distortions, revolt and reform in agricultural land relations* (Policy Research Working Paper 1164). World Bank, Washington.

Binswanger-Mkhize, H., & McCalla, A. F. (2010). The changing context and prospects for agricultural and rural development in Africa. In P. Pingali & R. Evenson (Eds.), *Handbook of agricultural economics*. Amsterdam: Elsevier.

Borras, S. M., Jr., Hall, R., Scoones, I., White, B., & Wolford, W. (2011). Towards a better understanding of global land grabbing: An editorial introduction. *Journal of Peasant Studies, 38*(2), 209–216.

Bravo-Ortega, C., & Lederman, D. (2009). La agricultura y el bienestar nacional en el mundo. *El Trimestre Economico, 76*(3), 577–617.

Brinkley, D. (1996). Bringing the Green Revolution to Africa: Jimmy Carter, Norman Borlaug, and the Global 2000 Campaign. *World Policy J., 13*(1), 53–62.

Bruinsma, J. (2009). The resources outlook: By how much do land, water and crop yields need to increase by 2050? In P. Conforti (Ed.), *Looking ahead in world food and agriculture: Perspectives to 2050*. Rome: Food and Agriculture Organization.

Butzer, R., Mundlak, Y., & Larson, D. F. (2003). Intersectoral migration in Southeast Asia: Evidence from Indonesia, Thailand, and the Philippines. *Journal of Agricultural and Applied Economics, 35*(Supplement), S105–S117.

Byerlee, D., Masters, W. A., & Robinson, D. (2017). From land grab to land development: Unlocking the potential for private investment to support small-holder farms in frontier areas. In P. Pengali & G. Feder (Eds.), *Agriculture and rural development in a globalizing world: Challenges and opportunities*. London and New York: Routledge.

Byiringiro, F., & Reardon, T. (1996). Farm productivity in Rwanda: Effects of farm size, erosion, and soil conservation investments. *Agricultural Economics, 15*(2), 127–136.

Carletto, C., Savastano, S., & Zezza, A. (2013). Fact or artifact: The impact of measurement errors on the farm size–productivity relationship. *Journal of Development Economics, 103*, 254–261.

Carter, M. R. (1984). Identification of the inverse relationship between farm size and productivity: An empirical analysis of peasant agricultural production. *Oxford Economic Papers, 36*(1), 131–145.

Christiaensen, L., Demery, L., & Kuhl, J. (2011). The (evolving) role of agriculture in poverty reduction: An empirical perspective. *Journal of Development Economics, 96*(2), 239–254.

Cobo, J. G., Dercon, G., & Cadisch, G. (2010). Nutrient balances in African land use systems across different spatial scales: A review of approaches, challenges and progress. *Agriculture, Ecosystems & Environment, 136*(1), 1–15.

Coeymans, J. E., & Mundlak, Y. (1992). Endogenous technology and sectoral productivity: Chile 1962–1982. *The Journal of Productivity Analysis, 3*(3), 257–275.

Collier, P., & Venables, A. J. (2012). *Land deals in Africa: Pioneers and Speculators* (Ox-Carre Research Paper 69). Oxford: Oxford University.

Collier, P., & Dercon, S. (2014). African agriculture in 50 years: Smallholders in a rapidly changing world? *World Development, 63*, 92–101.

Cornia, G. A. (1985). The social costs and benefits of biofuels: The intersection of environmental, energy and agricultural policy. *Applied Economic Perspectives and Policy, 13*(4), 513–534.

Crawford, E., Kelly, V., Jayne, T. S., & Howard, J. (2003). Input use and market development in Sub-Saharan Africa: An overview. *Food Policy, 28*(4), 277–292.

Dao, A., Sanou, J., Gracen, V., & Danquah, E. Y. (2015). Identifying farmers' preferences and constraints to maize production in two agroecological zones in Burkina Faso. *Agriculture Food Security, 4*(1), 1–7.

David, C. C., & Otsuka, K. (1994). *Modern rice technology and income distribution in Asia*. Boulder, CO: Lynne Rienner.

deGraft-Johnson, M., Suzuki, A., Sakurai, T., & Otsuka, K. (2016). 5 On the possibility of rice green revolution in rainfed areas in northern Ghana: An assessment of a management training program. In K. Otsuka & D. F. Larson (Eds.), *In pursuit of an African Green Revolution: Views from rice, maize farmers fields*. Tokyo: Springer.

Deininger, K., & Byerlee, D. (2012). The rise of large farms in land abundant countries: Do they have a future? *World Development, 40*(4), 701–714.

Deininger, K., Hilhorst, T., & Songwe, V. (2014). Identifying and addressing land governance constraints to support intensification and land market operation: Evidence from 10 African countries. *Food Policy, 48*, 76–87.

Deininger, K., Byerlee, D., Lindsay, J., Norton, A., Selod, H., & Stickler, M. (2011). *Rising global interest in Farmland: Can it yield sustainable and equitable benefits?*. Washington: World Bank.

Delgado, C., Narrod, C., & Tiongco, M. (2008). *Determinants and implications of the growing scale of livestock farms in four fast-growing developing countries* (Research Report 157). Washington: International Food Policy Research Institute.

Dercon, S. (2013). Agriculture and development: Revisiting the policy narratives. *Agricultural Economics, 44*(S1), 183–187.

Dethier, J. J., & Effenberger, A. (2012). Agriculture and development: A brief review of the literature. *Economic Systems, 36*(2), 175–205.

Deugd, M., Röling, N., & Smaling, E. M. A. (1998). A new praxeology for integrated nutrient management, facilitating innovation with and by farmers. *Agriculture, Ecosystems & Environment, 71*(1–3), 269–283.

Diagne, A., Midingoyi, S. K. G., & Kinkingninhoun-Medagbe, F. M. (2013). Impact of NERICA adoption on rice yield: Evidence from West Africa. In K. Otsuka & D. F. Larson (Eds.), *An African Green Revolution*. Dordrecht: Springer.

Diao, X., Hazell, P. B. R., & Thurlow, J. (2010). The role of agriculture in African development. *World Development, 38*(10), 1375–1383.

Djurfeldt, A. A., & Djurfeldt, G. (2013). Structural transformation and African smallholders: Drivers of mobility within and between the farm and non-farm sectors for eight countries. *Oxford Development Studies, 41*(3), 281–306.

Dorosh, P., & Thurlow, J. (2014). Can cities or towns drive African development? Economy wide analysis for Ethiopia and Uganda. *World Development, 63,* 113–123.

Drechsel, P., Gyiele, L., Kunze, D., & Cofie, O. (2001). Population density, soil nutrient depletion, and economic growth in Sub-Saharan Africa. *Ecological Economics, 38*(2), 251–258.

Drèze, J., & Sen, A. K. (1989). *Hunger and public action*. Oxford: Oxford University Press.

Duflo, E., Kremer, M., & Robinson, J. (2008). How high are rates of return to fertilizer? Evidence from field experiments in Kenya. *American Economic Review, 98*(2), 482–488.

Dzanku, F. M., Jirström, M., & Marstorp, H. (2015). Yield gap-based poverty gaps in rural sub-Saharan Africa. *World Development, 67,* 336–362.

Eastwood, R., Lipton, M., & Newell, A. (2010). Farm size. In P. L. Pingali & R. E. Evenson (Eds.), *Handbook of agricultural economics*. Elsevier, Amsterdam: North Holland.

Eicher, C. K., & Baker, D. C. (1982). *Research on agricultural development in Sub-Saharan Africa: a critical survey* (MSU International Development Paper 1). East Lansing: Michigan State University.

Estudillo, J. P., & Otsuka, K. (2013). Lessons from the Asian Green Revolution in Rice. In K. Otsuka & D. F. Larson (Eds.), *An African Green Revolution: Finding ways to boost productivity on small farms*. Dordrecht: Springer.

Evenson, R. E., & Gollin, D. (2003a). Assessing the impact of the Green Revolution, 1960–2000. *Science, 300*(5620), 758–762.

Evenson, R. E., & Gollin, D. (2003b). Crop genetic improvement in developing countries: Overview and summary. In R. E. Evenson & D. Gollin (Eds.), *Crop variety improvement and its effect on productivity: The impact of international agricultural research*. Cambridge, MA: CABI Publishing.

Evenson, R. E., & Gollin, D. (2003c). *Crop variety improvement and its effect on productivity: The impact of international agricultural research*. Cambridge, MA: CABI Publishing.

Fan, S., & Chan-Kang, C. (2005). Is small beautiful? Farm size, productivity, and poverty in Asian agriculture. *Agricultural Economics, 32*(s1), 135–146.

FAO. (2017). *Rural Africa in motion. Dynamics and drivers of migration South of the Sahara*. Rome: Food and Agriculture Organization. http://www.fao.org/3/I7951EN/i7951en.pdf.

FAOSTAT. (2015). *Food and agriculture organization corporate statistical database*. Rome: Food and Agriculture Organization.

Feder, G. (1985). The relation between farm size and farm productivity: The role of family labor, supervision and credit constraints. *Journal of Development Economics, 18*(2), 297–313.

Fischer, G. (2009). *How can climate change and the development of bioenergy alter the long-term outlook for food and agriculture?* Rome: Food and Agriculture Organization. Available at: http://www.fao.org/3/a-ak972e.pdf.

Fox, L., Haines, C., Muñoz, J. H., & Thomas, A. H. (2013). *Africa's got work to do: Employment prospects in the new century* (IMF Working Paper 13/201). Washington: International Monetary Fund.

Frelat, R., Lopez-Ridaura, S., Giller, K. E., Herrero, M., Douxchamps, S., Djurfeldt, A. A., et al. (2016). Drivers of household food availability in Sub-Saharan Africa based on big data from small farms. *Proceedings of National Academy of Sciences, 113*(2), 458–463.

Gardner, B. L. (2000). Economic growth and low incomes in agriculture. *American Journal of Agricultural Economics, 82*(5), 1059–1074.

Godfray, H. C. J., Beddington, J. R., Crute, I. R., Haddad, L., Lawrence, D., Muir, J. F., et al. (2010a). Food security: The challenge of feeding 9 billion people. *Science, 327*(5967), 812–818.

Godfray, H. C. J., Crute, I. R., Haddad, L., Lawrence, D., Muir, J. F., Nisbett, N., et al. (2010b). The future of the global food system. *Philosophical Transactions of the Royal Society of London. Series B, Biological sciences, 365*(1554), 2769–2777.

de Gorter, H., & Just, D. R. (2010). The social costs and benefits of biofuels: The intersection of environmental, energy and agricultural policy. *Applied Economic Perspectives and Policy, 32*(1), 4–32.

de Groote, H., Siambi, M., Friesen, D., & Diallo, A. (2002). Identifying farmers' preferences for new maize varieties in eastern Africa. In M. R. Bellon & J. R. Reeves (Eds.), *Quantitative analysis of data from participatory methods in plant breeding*. Mexico City: CIMMYT.

Haggblade, S., & Hazell, P. B. R. (2010). *Successes in African agriculture: Lessons for the future*. Baltimore: The Johns Hopkins University Press.

Haggblade, S., Hazell, P. B. R., & Reardon, T. (2010). The rural non-farm economy: Prospects for growth and poverty reduction. *World Development, 38*(10), 1429–1441.

Hall, A. J., & Richards, R. A. (2013). Prognosis for genetic improvement of yield potential and water-limited yield of major grain crops. *Field Crops Res., 143,* 18–33.

Hawkesford, M. J., Araus, J. L., Park, R., Calderini, D., Miralles, D., Shen, T., et al. (2013). Prospects of doubling global wheat yields. *Food Energy Security, 2*(1), 34–48.

Hayami, Y., & Kikuchi, M. (2000). *A rice village saga: Three decades of Green, Revolution in the Philippines*. London: Macmillan Press Ltd.

Hayami, Y., Quisumbing, A. R., & Adriano, L. (1990). *Towards an alternative land reform paradigm: A Philippine, perspective*. Manila: Ateneo de Manila University Press.

Hazell, P. B. R. (2009). Transforming agriculture: The Green Revolution in Asia. In D. Spielman & R. Pandya-Lorch (Eds.), *Millions Fed: Proven successes in agricultural development*. Washington: International Food Policy Research Institute.

Hazell, P. B. R., Poulton, C., Wiggins, S., & Dorward, A. (2010). The future of small farms: Trajectories and policy priorities. *World Development, 38*(10), 1349–1361.

Hossain, M. (1977). Farm size, tenancy and land productivity: An analysis of farm level data in Bangladesh agriculture. *Bangladesh Development Studies, 5*(3), 285–348.

Howard, J., Crawford, E., Kelly, V., Demeke, M., & Jeje, J. J. (2003). Promoting high-input maize technologies in Africa: The Sasakawa-Global 2000 experience in Ethiopia and Mozambique. *Food Policy, 28*(4), 335–348.

IPCC. (2014). *Intergovernmental panel on climate change. Climate change 2014: Impacts, adaptation, and vulnerability*. Available on the internet at: https://www.ipcc.ch/report/ar5/wg2/.

Irz, X., Lin, L., Thirtle, C., & Wiggins, S. (2001). Agricultural productivity growth and poverty alleviation. *Development Policy Review, 19*(4), 449–466.

de Janvry, A., & Sadoulet, E. (2010). Agricultural growth and poverty reduction: Additional evidence. *The World Bank Research Observer, 25*(1), 1–20.

Jayne, T., & Rashid, S. (2013). Input subsidy programs in Sub-Saharan Africa: A synthesis of recent evidence. *Agricultural Economics, 44*(6), 547–562.

Jeon, Y. D., & Kim, Y. Y. (2000). Land reform, income redistribution, and agricultural production in Korea. *Economic Development and Cultural Change, 48*(2), 253–268.

Jayne, T. S., Chamberlin, J., & Headey, D. D. (2014). Land pressures, the evolution of farming systems, and development strategies in Africa: A synthesis. *Food Policy, 48,* 1–17.

Jayne, T. S., Govereh, J., Wanzala, M., & Demeke, M. (2003). Fertilizer market development: A comparative analysis of Ethiopia, Kenya, and Zambia. *Food Policy, 28*(4), 293–316.

Johnson, M., Hazell, P., & Gulati, A. (2003). The role of intermediate factor markets in Asia's Green Revolution: Lessons for Africa? *American Journal of Agricultural Economics, 85*(5), 1211–1216.

Johnston, B. F., & Cownie, J. (1969). The seed-fertilizer revolution and labor force absorption. *American Economic Review, 59*(4), 569–582.

Kajisa, K. (2016). On the determinants of low productivity of rice farming in Mozambique: Pathways to intensification. In K. Otsuka & D. F. Larson (Eds.), *In pursuit of an African Green Revolution: Views from rice, maize farmers' fields*. Tokyo: Springer.

Kajisa, K., & Palanichamy, N. V. (2013). Chemical fertilizer, organic fertilizer, and cereal yields in India. In K. Otsuka & D. F. Larson (Eds.), *An African Green Revolution: Finding ways to boost productivity on small farms*. Dordrecht: Springer.

Khonje, M., Manda, J., Alene, A. D., & Kassie, M. (2015). Analysis of adoption and impacts of improved maize varieties in Eastern Zambia. *World Development, 66,* 695–706.

Khush, G. S. (2013). Strategies for increasing the yield potential of cereals: Case of rice as an example. *Plant Breeding, 132*(5), 433–436.

Kijima, Y. (2016). On the possibility of Rice Green Revolution in rainfed areas in Uganda: Impact evaluation of a management training program and guidebook distribution. In K. Otsuka & D. F. Larson (Eds.), *In pursuit of an African Green Revolution: Views from rice and maize farmers' fields*. Tokyo: Springer.

Kimhi, A. (2006). Plot size and maize productivity in Zambia: Is there an inverse relationship? *Agricultural Economics, 35*(1), 1–9.

Kostandini, G., Rovere, R. L., & Abdoulaye, T. (2013). Potential impacts of increasing average yields and reducing maize yield variability in Africa. *Food Policy, 43,* 213–226.

Kumar, S. (1979). *Impact of subsidised rice on food consumption and nutrition in Kerala Research Report 5*. Washington: International Food Policy Research Institute.

Kutcher, G., & Scandizzo, P. L. (1981). *The agricultural economy of North-East Brazil*. Baltimore: World Bank and Johns Hopkins University Press.

Lamb, R. L. (2003). Inverse productivity: Land quality, labor markets, and measurement error. *Journal of Development Economics, 71*(1), 71–95.

Langyintuo, A. S., Mwangi, W., Diallo, A. O., MacRobet, J., Dixon, J., & Bänziger, M. (2010). Challenges of the maize seed industry in eastern and southern Africa: A compelling case for private-public intervention to promote growth. *Food Policy, 35*(4), 323–331.

Larson, D. F., & Mundlak, Y. (1997). On the intersectoral migration of agricultural labor. *Economic Development and Cultural Change, 45*(2), 295–319.

Larson, D. F., & Leon, M. (2006). How endowments, accumulations, and choice determine the geography of agricultural productivity in Ecuador. *The World Bank Economic Review, 20*(3), 449–471.

Larson, D. F., & Gurara, D. Z. (2013). *A conceptual model of incomplete markets and the consequences for technology adoption policies in Ethiopia* (Policy Research Working Paper 6681). Washington: World Bank.

Larson, D. F., Anderson, J. R., & Varangis, P. (2004). Policies on managing risk in agricultural markets. *The World Bank Research Observer, 19*(2), 199–230.

Larson, D. F., Otsuka, K., Matsumoto, T., & Kilic, T. (2014). Should African rural development strategies depend on smallholder farms? An exploration of the inverse-productivity hypothesis. *Agricultural Economics, 45*(3), 355–367.

Larson, D. F., Savastano, S., Murray, S., & Palacios-Lopez, A. (2016). Are women less productive farmers? How markets and risk affect fertilizer use, productivity, and measured gender effects in

Uganda. In K. Otsuka & D. F. Larson (Eds.), *In pursuit of an African Green Revolution: Views from rice and maize farmers' fields*. Tokyo: Springer.
Lipton, M. (2006). Can small farmers survive, prosper, or be the key channel to cut mass poverty. Electron. *Journal of Development and Agricultural Economics, 3*(1), 58–85.
Lipton, M. (2009). *Land reform in developing countries: Property rights and property wrongs*. London: Routledge.
Lipton, M. (2012). *Learning from others: Increasing agricultural productivity for human development in Sub-Saharan Africa United Nations Development Programme Regional Bureau for Africa* (Working Paper 2012-007). New York: United Nations.
Lobell, D. B., Cassman, K. G., & Field, C. B. (2009). Crop yield gaps: Their importance, magnitudes and causes. *Annual Review of Environment and Resources, 34*(1), 179–204.
Lowder, S. K., Skoet, J., & Singh, S. (2014). *What do we really know about the number and distribution of farms and family farms in the world?* (ESA Working Paper 14-02). Rome: Food and Agriculture Organization.
Lowder, S. K., Skoet, J., & Raney, T. (2016). The number, size, and distribution of farms, smallholder farms, and family farms worldwide. *World Development, 87,* 16–29.
Lunduka, R., Fisher, M., & Snapp, S. (2012). Could farmer interest in a diversity of seed attributes explain adoption plateaus for modern maize varieties in Malawi? *Food Policy, 37*(5), 504–510.
Marenya, P., & Barrett, C. (2007). Household-level determinants of adoption of improved natural resources management practices among smallholder farmers in western Kenya. *Food Policy, 32*(4), 515–536.
Marenya, P., & Barrett, C. (2009). State-conditional fertilizer yield response on Western Kenyan farms. *American Journal of Agricultural Economics, 91*(4), 991–1006.
Masters, W. A., Djurfeldt, A. A., De Haan, C., Hazell, P., Jayne, T., Jirström, M., et al. (2013). Urbanization and farm size in Asia and Africa: Implications for food security and agricultural research. *Global Food Security, 2*(3), 156–165.
Mathenge, M. K., Smale, M., & Olwande, J. (2014). The impacts of hybrid maize seed on the welfare of farming households in Kenya. *Food Policy, 44,* 262–271.
Matsumoto, T., & Yamano, T. (2013). Maize, soil fertility, and the Green Revolution in East Africa. In K. Otsuka & D. F. Larson (Eds.), *An African Green Revolution: Finding ways to boost productivity on small farms*. Dordrecht: Springer.
Maxwell, S., & Slater, R. (2003). Food policy old and new. *Development Policy Review, 21*(5–6), 531–553.
McDermott, J. J., Staal, S. J., Freeman, H. A., Herrero, M., & Van de Steeg, J. A. (2010). Sustaining intensification of smallholder livestock systems in the tropics. *Livestock Science, 130*(1), 95–109.
Meadows, D. H., Meadows, D. L., Randes, J., & Behrens, W. W. (1972). *The limits to growth*. New York: Universe Book.
Morris, M., Kelly, V., Kopicki, R., & Byerlee, D. (2007). *Fertilizer use in African Agriculture: Lessons learned and good practice guidelines*. Washington, D.C.: World Bank.
Moseley, W. G., Carney, J., & Becker, L. (2010). Neoliberal policy, rural livelihoods, and urban food security in West Africa: A comparative study of The Gambia, Côte d' Ivoire, and Mali. *Proceedings of National Academy of Sciences, 107*(13), 5774–5779.
Mueller, N. D., Gerber, J. S., Johnston, M., Ray, D. K., Ramankutty, N., & Foley, J. A. (2012). Closing yield gaps through nutrient and water management. *Nature, 490*(7419), 254–257.
Mundlak, Y. (1988). Endogenous technology and the measurement of productivity. In S. M. Capalbo & J. M. Antle (Eds.), *Agricultural productivity: Measurement and explanation*. Washington: Resources for the Future.
Mundlak, Y., Cavallo, D., & Domenech, R. (1989). *Agriculture and economic growth in Argentina, 1913-84* (Research Report 76). Washington: International Food Policy Research Institute.
Mundlak, Y., Butzer, R., & Larson, D. F. (2012). Heterogeneous technology and panel data: The case of the agricultural production function. *Journal of Development Economics, 99*(1), 139–149.
Muraoka, R., Matsumoto, T., Jin, S., & Otsuka, K. (2016). On the possibility of a maize green revolution in the highlands of Kenya: An assessment of emerging intensive farming systems. In

K. Otsuka & D. F. Larson (Eds.), *In pursuit of an African Green Revolution: Views from rice and maize farmers' fields*. Tokyo: Springer.

Nakano, Y., Kajisa, K., & Otsuka, K. (2015). *To what extent does modified system of rice intensification (SRI) training increase productivity of small-scale cultivation in a rain-fed area? Evidence from Tanzania*. Tokyo, Japan: National Graduate Institute for Policy Studies, Mimeo,

Nakano, Y., Kajisa, K., & Otsuka, K. (2016). On the possibility of rice green revolution in irrigated and rainfed areas in Tanzania: An assessment of management training and credit programs. In K. Otsuka & D. F. Larson (Eds.), *In pursuit of an African Green Revolution: Views from rice and maize farmers' fields*. Tokyo: Springer.

Nakano, Y., Bamba, I., Diagne, A., Otsuka, K., & Kajisa, K. (2013). The possibility of a rice Green Revolution in large-scale irrigation schemes in Sub-Saharan Africa. In K. Otsuka & D. F. Larson (Eds.), *An African Green Revolution: Finding ways to boost productivity on small farms*. Dordrecht: Springer.

Nega, B., Adenew, B., & Gebre Sellasie, S. (2003). Current Land policy issues Ethiop. *Land Reform, Land Settlement and Cooperatives, 11*(3), 103–124.

Neumann, K., Verburg, P. H., Stehfest, E., & Müller, C. (2010). The yield gap of global grain production: A spatial analysis. *Agricultural Systems, 103*(5), 316–326.

Njeru, T. N., Mano, Y., Otsuka, K. (2016, forthcoming). Role of access to credit in rice production in Sub Saharan Africa: The case of Mwea irrigation scheme in Kenya. *Journal of African Economies, 25*(2), 300–321.

Nkonya, E., Pender, J., Jagger, P., Sserunkuuma, R., Kaissi, C., & Ssali, H. (2004). *Strategies for sustainable land management and poverty reduction in Uganda Research Report 133*. Washington: International Food Policy Research Institute.

Norgrove, L., & Hauser, S. (2016). Biophysical criteria used by farmers for fallow selection in West and Central Africa. *Ecological Indicators, 61*, 141–147.

Ojiem, J., Franke, A., Vanlauwe, B., de Ridder, N., & Giller, K. (2014). Benefits of legume–maize rotations: Assessing the impact of diversity on the productivity of smallholders in Western Kenya. *Field Crops Research, 168*, 75–85.

Otsuka, K., & Yamano, T. (2005). The possibility of a Green Revolution in Sub-Saharan Africa: evidence from Kenya. *Journal of Development and Agricultural Economics, 2*(1), 7–19.

Otsuka, K., & Yamano, T. (2006). Introduction to the special issue on the role of non-farm income in poverty reduction: evidence from Asia, East Africa. *Agricultural Economics, 35*(s3), 393–397.

Otsuka, K., & Larson, D. F. (2013). *An African Green Revolution: Finding ways to boost productivity on small farms*. Dordrecht: Springer.

Otsuka, K., & Larson, D. F. (2016). *In pursuit of an African Green Revolution: Views from rice and maize farmers' fields*. Tokyo: Springer.

Otsuka, K., Liu, Y., & Yamauchi, F. (2016). Growing advantage of large farms in Asia and its implications for global food security. *Global Food Security, 11*, 5–10.

Otsuka, K., & Place, F. (2015). Land tenure and agricultural intensification in Sub-Saharan Africa. In C. Monga & J. Y. Lin (Eds.), *The Oxford handbook of Africa and economics: Volume 2: Policies and practices*. Oxford: Oxford University Press.

Pardey, P. G., Beintema, N., Dehmer, S., & Wood, S. (2006). *Agricultural research: A growing global divide?*. Washington: International Food Policy Research Institute.

de Paul Obade, V., & Lal, R. (2013). Assessing land cover and soil quality by remote sensing and geographical information systems (GIS). *CATENA, 104*, 77–92.

Pingali, P. L. (2012). Green Revolution: Impacts, limits, and the path ahead. *Proceedings of National Academy of Sciences, 109*(31), 12302–12308.

Pingali, P. L., & Rosegrant, M. W. (1994). *Confronting the environmental consequences of the Green Revolution in Asia* (EPDT Discussion Paper 2). Washington: International Food Policy Research Institute.

Pinstrup-Andersen, P., & Hazell, P. B. (1985). The impact of the Green Revolution and prospects for the future. *Food Reviews International, 1*(1), 1–25.

Place, F., Barrett, C., Freeman, H. A., Ramisch, J. J., & Vanlauwe, B. (2003). Prospects for integrated soil fertility management using organic and inorganic inputs: Evidence from smallholder African agricultural systems. *Food Policy, 28*(4), 365–378.

Popkin, B. M. (1998). The nutrition transition and its health implications in lower-income countries. *Public Health Nutrition, 1*(01), 5–21.

Poulton, C., Kydd, J., & Dorward, A. (2006). Overcoming market constraints on pro-poor agricultural growth in Sub-Saharan Africa. *Development Policy Review, 24*(3), 243–277.

Ray, D. K., Mueller, N. D., West, P. C., & Foley, J. A. (2013). Yield trends are insufficient to double global crop production by 2050. *PLoS ONE, 8*(6), e66428.

Ray, D. K., Ramankutty, N., Mueller, N. D., West, P. C., & Foley, J. A. (2012). Recent patterns of crop yield growth and stagnation. *Nature Communications, 3*, 1297. https://doi.org/10.1038/ncomms2296.

Reij, C. P., & Smaling, E. M. A. (2008). Analyzing successes in agriculture and land management in Sub-Saharan Africa: Is macro-level gloom obscuring positive micro-level change? *Land Use Policy, 25*(3), 410–420.

Ricker-Gilbert, J., Mason, N. M., Darko, F. A., & Tembo, S. T. (2013). What are the effects of input subsidy programs on maize prices? Evidence from Malawi and Zambia. *Agricultural Economics, 44*(6), 671–686.

Rosegrant, M. W., & Hazell, P. B. R. (2000). *Transforming the Rural Asian Economy: The Unfinished Revolution*. Hong Kong: Oxford University Press.

Sakurai, T. (2016). On the determinants of high productivity in rice farming in irrigated areas in Senegal: Efficiency of large-scale vs. small-scale irrigation schemes. In K. Otsuka & D. F. Larson (Eds.), *In pursuit of an African Green Revolution: Views from rice and maize farmers' fields*. Tokyo: Springer.

Satterthwaite, D., McGranahan, G., & Tacoli, C. (2010). Urbanization and its implications for food and farming. *Philosophical Transactions of the Royal Society of London. Series B, Biological Sciences, 365*(1554), 2809–2820.

Searchinger, T. D., Estes, L., Thornton, P. K., Beringer, T., Notenbaert, A., Rubenstein, D., et al. (2015). High carbon and biodiversity costs from converting Africa's wet savannahs to cropland. *Nature Climate Change, 5*(5), 481–486.

Singh, B. (1985). *Agrarian structure*. Agricole, New Delhi: Technological Change and Poverty.

Sen, A. K. (1966). Peasants and dualism with or without surplus labor. *Journal of Political Economy, 74*(5), 425–450.

Sheahan, M., Black, R., & Jayne, T. (2013). Are Kenyan farmers underutilizing fertilizer? Implications for input intensification strategies and research. *Food Policy, 41*, 39–52.

Smale, M. (1994). Maize is life: Malawi's delayed Green Revolution. *World Development, 23*(5), 819–831.

Smale, M., & Jayne, T. S. (2010). Seeds of success in retrospect: Hybrid maize in Eastern and Southern Africa. In S. Haggblade & P. B. R. Hazell (Eds.), *Successes in African Agriculture: Lessons for the future*. Washington: International Food Policy Research Institute.

Smale, M., Byerlee, D., & Jayne, T. (2013). Maize revolution in Sub-Saharan Africa. In K. Otsuka & D. F. Larson (Eds.), *An African Green Revolution: Finding ways to boost productivity on small farms*. Dordrecht: Springer.

Smale, M., Moursi, M., & Birol, E. (2015). How does adopting hybrid maize affect dietary diversity on family farms? Micro-evidence from Zambia. *Food Policy, 52*, 44–53.

Snapp, S., Blackie, M., Gilbert, R., Bezner-Kerr, R., & Kanyama-Phiri, G. (2010). Biodiversity can support a Greener Revolution in African. *Proceedings National Academy of Sciences, 107*(48), 20840–20845.

Srinivasan, T. N. (1972). Farm size and productivity implications of choice under uncertainty. Sankhyā: Indian. *Journal of the Royal Statistical Society: Series B, 34*(4), 409–420.

Stevenson, J. R., Villoria, N., Byerlee, D., Kelley, T., & Maredia, M. (2013). Green Revolution research saved an estimated 18 to 27 million hectares from being brought into agricultural production. *Proceedings of National Academy of Sciences, 110*(21), 8363–8368.

Suri, T. (2011). Selection and comparative advantage in technology adoption. *Econometorica, 79*(1), 159–209.

Thornton, P. K. (2010). Livestock production: Recent trends future prospects. *Philosophical Transactions of the Royal Society of London. Series B, Biological sciences, 365*(1554), 2853–2867.

Timmer, C. P. (1997). Building efficiency in agricultural marketing: The long-run role of BULOG in the Indonesian food economy. *Journal of International Development, 9*(1), 133–146.

Tittonell, P., & Giller, K. E. (2013). When yield gaps are poverty traps: The paradigm of ecological intensification in African smallholder agriculture. *Field Crops Res., 143,* 76–90.

Tripp, R. (1993). Invisible hands, indigenous knowledge and inevitable fads: Challenges to public sector agricultural research in Ghana. *World Development, 21*(12), 2003–2016.

United Nations. (2018). *World urbanization prospects, The 2018 Revision*. New York: Department of Economic and Social Affairs, United Nations.

Vollrath, D. (2007). Land distribution and international agricultural productivity. *American Journal of Agricultural Economics, 89*(1), 202–216.

van Zyl, J., Binswanger, H., & Thirtle, C., (1995). *The relationship between farm size and efficiency in South African Agriculture* (Policy Research Working Paper 1548). Washington: World Bank.

Wiggins, S., Kirsten, J., & Llambí, L. (2010). The future of small farms. *World Development, 38*(10), 1341–1348.

Winter-Nelson, A., & Temu, A. (2005). Impacts of prices and transactions costs on input usage in a liberalizing economy: Evidence from Tanzanian coffee growers. *Agricultural Economics, 33*(3), 243–253.

World Bank. (2015). *Commodity markets outlook October 2015*. Washington: World Bank.

World Bank Pink Sheet. (2015). *Global economic monitor commodities*. Available at: http://databank.worldbank.org/data/databases/commodity-price-data.

Xu, Z., Guan, Z., Jayne, T. S., & Black, R. (2009). Profitability of applying fertilizer on maize for smallholder farmers in Zambia. *Agricultural Economics, 40*(4), 437–446.

Yamauchi, F., & Larson, D. F. (2016). *Long-term impacts of an unanticipated risk event: The 2007/08 food price crisis and child growth in Indonesia* (World Bank Policy Research Working Paper 7627). Washington: World Bank.

Yotopoulos, P. A., & Lau, L. J. (1973). A test for relative economic efficiency: Some further results. *American Economic Review, 63*(1), 214–223.

Zeng, D., Alwang, J., Norton, G. W., Shiferaw, B., Jaleta, M., & Yirga, C. (2015). Ex post impacts of improved maize varieties on poverty in rural Ethiopia. *Agricultural Economics, 46*(4), 1–12.

Zerfu, D., & Larson, D. F. (2010). *Incomplete markets and fertilizer use: Evidence from Ethiopia* (World Bank Policy Research Working Paper Series 5235). Washington: World Bank.

Zoomers, A. (2010). Globalisation, the foreignisation of space: Seven processes driving the current global land grab. *Journal of Peasant Studies, 37*(2), 429–447.

10

Policies for Improved Food Security: The Roles of Land Tenure Policies and Land Markets

Stein T. Holden

Abstract This chapter provides an overview of what we know about farm size distributions, the emerging land markets, the role of tenure systems, tenure reforms and land policies in shaping the distribution of increasingly scarce land resources. The primary focus is on Africa while making some comparisons with Asia. Climate risk and change have serious implications for household vulnerability and food security. While there is a need to absorb further population growth in rural areas, a rapid rise in rural–urban migration is inevitable. Careful land use planning and tenure reforms are needed to smooth the transition towards more intensive land use.

1 Introduction

Land scarcity is growing in many parts of Africa, and land markets are emerging and becoming more active (Holden et al. 2008). Population growth has also put increasing pressures on customary tenure systems. The sharp increase in demand for land following the hikes in food and energy prices in 2007–2010 revealed weaknesses in land tenure systems and policies (Deininger et al. 2014). The increased investor demand for land created fears that vulnerable groups and smallholders would lose their land rights and become more food insecure (de Schutter 2011; German et al. 2013). This chapter provides an overview of what we know about the emerging land markets in Africa, the role of tenure systems, tenure reforms and land policies in shaping the distribution of increasingly scarce land resources and draws implications for future farm size distributions and livelihood opportunities for the current large and growing rural populations in the continent. Rural–urban migration and international migration are expected to grow rapidly, and management of this flow of people will have to be a central element of more holistic tenure policies. Provision of food at affordable prices, for the growing rural and urban populations facing climate change with more turbulent weather conditions, is an increasing challenge.

S. T. Holden (✉)
School of Economics and Business, Centre for Land Tenure Studies, Norwegian University of Life Sciences, P. O. Box 5003, 1432 Ås, Norway
e-mail: stein.holden@nmbu.no

There are signs of a more rapid transformation of the agricultural sector in many African countries with high economic growth, infrastructure investments and new technologies. New technologies have also sharply reduced the costs of formalisation of land rights. Land tenure reforms using low-cost methods have been scaled up in a number of countries and appear to have contributed to enhanced tenure security and better functioning land markets. However, the devil is in the detail of land tenure reforms, in terms of specific details, interpretation and implementation. These can have long-term impacts on the distribution of power, land resources, economic performance, welfare and welfare distribution.

Reforms that enhance land rental markets appear important as a tool to facilitate a transformation towards more productive commercial agriculture that is capable of feeding the growing urban populations in Africa. The rapid urbanisation process in some areas is putting high pressure on tenure systems, and good governance is extremely important to minimise conflicts and facilitate smooth transition. The recent political unrest in Ethiopia was also partly triggered by the need to expand the borders of Addis Ababa, met by protests by those losing their land with minimal compensation offered compared to the potential value of the land. Rapid economic transformation and shocks also cause rapid social changes and stress that can ignite social unrest and political conflicts, threatening political stability.

I aim to assess the following research questions: What are the implications of population growth and increasing land scarcity for livelihood opportunities and food security in Africa? Will land markets stimulate land access for the land-poor? Will tenure reforms benefit small farmers or be to their disadvantage?

2 Population Growth, Farm Size Distribution and Food Security

According to the High Level Panel of Experts on Food Security and Nutrition (HLPE 2013), there was a general decrease in average farm size in Africa and in China during the period 1930–2000. Using data from 2000 for 81 countries, they find that 73% of the farms were smaller than 1 ha and 85% were smaller than 2 ha. In 14 African countries, 80% of the farms were smaller than 2 ha, and these utilised 25% of the agricultural land. Masters et al. (2013) indicates that the trends towards smaller holdings have changed in Asia in recent years. Jayne et al. (2014) find that farm sizes are shrinking in land-constrained countries in Africa, such as Ethiopia, Kenya, Malawi, Nigeria, Rwanda and Uganda, but are increasing in some more land-abundant countries, such as Tanzania and Zambia in recent years.

In low-income countries, we may distinguish two main trends in farm size distribution. The first trend, which is affecting the majority of densely populated rural areas, is an increasing pressure on land due to population growth and absorption of more people on the land. This trend is strongest in sub-Saharan Africa (SSA), where it is predicted that the rural population will increase by 60% from 2015 to 2050 (United

Nations 2014). This leads to continued fragmentation into smaller farm sizes and is partly driven by the dominant inheritance rules, where land tends to be split among children of the parents.

How far this fragmentation process will go in different environments will depend on what is a minimum farm size for an acceptable rural livelihood, given the combination of farm productivity and alternative sources of income that are available to those living on the farm. What is an acceptable rural livelihood depends on cultural norms, as well as alternative and complementary economic opportunities. A rural livelihood may potentially become a poverty trap ('small is ugly'); if population growth in such areas is stronger than productivity growth, the costs of escaping the locality are too high, and the known alternatives elsewhere are not considered attractive.

Severe shocks to such a poverty-trapped vulnerable locality may cause desperate migration and/or local disaster. Covariate shocks are more serious in isolated areas because they are associated with large endogenous price changes that add to the costs of consumption smoothing. Households that may be net sellers of food under favourable climatic conditions may become net buyers of food in drought or flood years. Poor market integration causes loss-loss price effects for such households, as they have to sell their surpluses in good years at very low prices and buy their deficits at very high prices in years with covariate weather shocks. If they keep livestock as a means of protecting themselves against such shocks, livestock prices also tend to fluctuate in their disfavour, as they may have to sell livestock at low prices in order to buy food at a high price in shock years. When they have recovered after a shock and want to rebuild their livestock, livestock prices are again likely to be high (Holden and Shiferaw 2004).

The trend towards fragmentation into smaller farms is also associated with an increasing share of rural farm households becoming deficit producers of staple food. Climate change may contribute to larger risks and uncertainty in agricultural production and enhance the vulnerability of smallholder farmers, who may become even more food insecure. Investments in infrastructure and market integration may contribute to reduce price fluctuations due to covariate risk, and technological improvements may contribute to adaptation and reduction in production risks. However, provision of food to the rapidly growing urban populations will require large increases in agricultural production and in marketed surpluses.

There are likely to be diminishing returns to labour as labour/land ratios are increasing when farm sizes get very small. An inverse relationship (IR) between farm size and land productivity has frequently been observed within smallholder agriculture, and this relationship appears to be not only due to land quality differences and measurement error (Carletto et al. 2013). Labour and land market imperfections are at the heart of the IR. Moral hazard causes hired labour not to be a perfect substitute for family labour. In combination with seasonality and search costs in the labour market, there is a tendency that higher labour/land ratios on farms are associated with more intensive labour use per unit land, and this may also be associated with lower shadow wages and thus higher output per unit land (land use intensification). However, this may also be affected by access to other productivity-enhancing inputs than labour. If the most land-poor but labour-rich have limited access to input markets and improved technologies, or need access to credit in order to afford to buy inputs, they may be too poor to invest and to be efficient (Holden and Binswanger 1998). There is also a risk that their labour-intensive practices contribute more to soil mining or land degradation if they do not replace lost nutrients or conserve the land properly (Shiferaw and Holden 1998).

With continued rural population growth and expansion of super-small farms, push- rather than demand-driven outmigration is likely to increase over time, and employment creation for such migrants will become critical for economic development and social stability. The stochastic nature of shocks is also likely to cause stochastic migration flows, unless there is institutional and governance capacity to cushion the effects of such shocks.

A change in inheritance rules may help to prevent fragmentation into smaller farms, e.g. by requiring that only one of the children in the family can inherit the land to ensure that the farm size remains intact. Setting of minimum farm sizes for legal registration may be another way to attempt to stem this trend, but whether and how it will work is an empirical issue. Facilitation of consolidation into larger farm sizes is another option. The success of such policies will depend on the incentives and the degree of compliance. There are few studies that have assessed the costs and benefits of such land fragmentation. A study by Ali and Deininger (2015) in Rwanda—the most densely populated country in Africa, where the average farm size is 0.72 ha (on average split into four parcels)—shows that this is not sufficient to meet household food needs. The national land policy prohibits splitting of farms into units smaller than 1 ha and advocates land consolidation and mobilisation of land for investors. The study finds, however, a strong IR when controlling for household-specific shadow wages and a strong ability of small farms to absorb labour in a productive way, while the IR disappears when market wages are used to calculate area-based profits. A policy that generates employment and higher wages outside agriculture is likely to be better than a forced consolidation policy that may do more harm than good in such a context.

3 Expansion of Medium-Sized Farms

The other trend of significant importance is the expansion of medium-sized farms observed in, for example, Ghana, Kenya, Malawi and Zambia (Anseeuw et al. 2016; Sitko and Jayne 2014; Jayne et al. 2016). This trend is associated with urban expansion and land acquisition by African elites. The trend is strong enough to threaten small farm development and reduce land access for land-poor smallholders. This expansion may have been boosted by rising energy and food prices and facilitated by weak land governance and tenure systems. It is uncertain whether customary tenure reforms will hinder or further boost this development through privatisation of land rights and land market development. While some earlier studies in East Africa did not reveal that land sales contributed to a more skewed land distribution (Holden et al. 2008), more recent studies point in the direction of a more skewed land distribution because of the expansion of large and medium-sized farms in combination with further subdivision of small farms (Jayne et al. 2016).

Whether this increase in medium-sized farms will lead to increased land productivity and improved food security overall depends on a number of factors. If those taking over or buying such farms are doing it to establish productive agriculture and they possess the necessary skills, social networks and other forms of capital, this may enhance efficiency. If, on the other hand, they do it to establish a rural resort for retirement and as a tax shelter, the outcome may be reduced land productivity. Anseeuw et al. (2016) found that a large share of the owners of medium-sized farms in Malawi were urban dwellers and part-time farmers. Sitko and Jayne (2014) found a similar situation in Zambia. Their studies indicate that the rural growth linkages become weaker with this kind of change in ownership structure, as more of the income remains in urban areas.

There is a shortage of good studies carefully comparing productivity on medium-sized versus small farms. In general, there are few economies of scale in agriculture in low-income countries (Binswanger and Rosenzweig 1986). Anseeuw et al. (2016) find that the share of unutilised land increases with farm size in Malawi. While policies in Malawi favoured estate agriculture up until the early 1990s, including through prohibition of smallholder tobacco production, the removal of these policies caused a rapid expansion of smallholder tobacco production while many tobacco estates put more land under fallow. The relatively low utilisation of estate land in the country contributed to a 'market-assisted land redistribution' project, providing 15,000 landless or near landless households with farmland, and this project appears to have had positive impacts (Simtowe et al. 2013). On the other hand, Ali et al. (2015) find that medium-sized farms have higher land productivity than small farms in Ethiopia. Farms in the range of 10–20 ha had the highest land productivity for most crops. However, the data were not good enough to control for land quality or labour use intensity, which potentially could be correlated with farm size. They found that commercial farms had on average one permanent job per 20 ha, with some additional temporary jobs, and this is far below smallholder agriculture in the country where a family farm may only be 1 ha and thus possibly absorbing 20 times more labour per unit land. With about 80 million of the 100 million population in Ethiopia still engaged primarily in smallholder agriculture, and with population growth of about 2.6 million per year, providing alternative employment to transform smallholder agriculture to commercial agriculture is not a trivial challenge.

A conversion from small farms to larger (medium-sized) farms will require a livelihood solution for those populations that have to give up their small farms in order to provide the space for larger farms. If those populations can be provided with attractive better livelihoods in urban areas and the non-farm sector, the transformation of small farms into medium and large farms may be socially optimal and be consistent with what has happened in many western countries. It could lead to a drastic reduction in rural poverty. However, it would have to be driven by very strong economic growth outside the agricultural sector. It may be too optimistic to hope for this in all developing countries with large rural populations concentrated on small farms, although economic growth has increased in many African countries over the last two decades. On the other hand, it may be less challenging to establish medium and large farms in areas with low population density, although this will also depend on traditional land rights and the existence of customary tenure systems.

In China, rural wages have started to increase in recent years, and this has also triggered a move towards larger farm sizes there. However, it took many years of strong economic growth, even in China with its population control policy, to eliminate the surplus labour in rural areas to the extent that rural wages have started to rise due to labour scarcity in these areas. It may take even longer to see such an effect in countries with similar economic growth and larger rural population growth.

4 Increasing Landlessness and Migration

We are still haunted by Malthusian theory. There is a danger that poor smallholders become too poor to invest and unable to maintain the productivity of the land on which they subsist (Holden and Binswanger 1998; Shiferaw and Holden 1998), unless systematic actions are taken to promote sustainable intensification of smallholder agriculture. While there are many examples of success stories (Boserupian pathways), and evidence that promotion of tenure security enhances incentives to conserve land, secure tenure rights is only a necessary—not a sufficient—condition to ensure sustainable land use on smallholder farms (Holden et al. 1998, 2009). Comprehensive tailor-made policies are needed to handle diverse and densely populated rural livelihood systems. Whether diversification, intensification and growth can be facilitated within the economy, or whether outmigration or strict family planning to curb population growth is needed, is an empirical issue. China succeeded with the household responsibility system, population control and export-oriented industrialisation policies, and rural wages and farm sizes have started to rise in recent years. For more densely populated and land-constrained African countries where economic growth has picked up, rural population growth is still high and continues to concentrate on small farms. This represents a massive push towards the minimum bearable farm size, and a rapidly growing number of young people are pushed over the cliff into landlessness. Growing national and international youth migration is an inevitable outcome and a huge challenge.

In Asia, landlessness has in the past been closely associated with poverty, due to the importance of land as a source of income in rural areas (Pakistan: Anwar et al. 2004; India: Meenakshi and Ray 2002; Philippines: Balisacan 1993). However, landlessness may also be an indicator of growing diversification in rural economies, and the finding of poor correlation in some countries between poverty and landlessness shows that landlessness can also be a sign of better opportunities (Lanjouw 2007). Ravallion and van de Walle (2008) found that recent increases in landlessness in

Vietnam are not associated with higher levels of poverty. The differentiation in rural areas of Vietnam after strong economic growth appears to have reached a stage where landlessness is decoupled from poverty.

Landlessness is growing rapidly in densely populated poor agrarian economies, and in such economies, the landless may be among the poorest of the poor. In undifferentiated land-constrained rainfed agriculture, it may even be very hard to survive if landless as there may be very few jobs available that provide a livelihood throughout the year, unless one has some specialised skills or education. Outmigration may thus be the best and only option for the landless in such areas that have passed a certain threshold level of population density (Bezu and Holden 2014). While this may occur quite suddenly in some egalitarian societies with equitable land distribution, it may occur more gradually in areas with more variation in farm sizes. Climatic and other forms of covariate shocks may also trigger massive migration within short periods of time unless organised actions are taken to buffer such shocks. Migrants have to leave immobile assets, including their land, behind. Without compensation for such immobile assets and with inability to return, they are likely to be extremely poor and also to be very vulnerable wherever they go. Whether it is better that these asset values stay behind with the remaining rural population, or go with the migrants as capital to help them start a new livelihood, is also an empirical issue. Extreme rural poverty, contrasting with better-off urban livelihoods and adequate job opportunities where migrants go, may point in direction of retaining the asset values in the rural areas of origin. This implies that, for example, the child in the family that remains behind and takes over the family farm does not need to compensate their siblings who have to migrate and do not inherit any of the immobile assets.

On the other hand, with a functioning rural land sales market, the whole family may decide to sell their immobile assets. While this may help migrants to get start-up capital, there is also a risk that it can lead to distress land sales, increasing landlessness and accumulation of land in the hands of the wealthy (Holden et al. 2008; Jayne et al. 2016). In countries with powerful urban elites who influence land governance through law formation or law violations, there is a higher risk that land tenure reforms and land markets contribute to more inegalitarian land distribution, in a way that also makes poverty worse.

In many developing countries, the traditional way of preventing landlessness in rural areas has been through inheritance, sharing the land among the children. In some countries, there has even been a constitutional right for rural residents to access land for free as they become old enough to establish their own family (e.g. Ethiopia). Customary tenure systems have also had such rules for land allocation to clan members. Such allocation mechanisms function only up the point where land scarcity creates a gap between demand and supply of land. In Ethiopia, they have recently started to allocate rehabilitated communal lands to groups of landless youth. Providing them with temporary licences to extract mineral resources is another approach to help them generate starting capital for another business (Holden and Tilahun 2018).

5 Land Market Development

Land scarcity and the existence of a price for land, where there is a willing seller and a willing buyer, are the fundamental requirements for a land sales transaction and a land market to occur. Such land sales transactions can lead to more productive land use, if the buyer primarily aims to use the land for productive purposes. However, if they were able to buy it under distress conditions for the seller, and thus obtained it at a very favourable price, the desperate seller may under the circumstances have been forced to sell the land at a price below their own long-term value of the land. It is thus not necessarily the case that land sales transactions transfer land from less productive to more productive farmers. But under normal conditions, where the purchase is for productive purposes, buyers of land may also turn out to be more productive users of the land than those who sold the land. However, output price fluctuations, such as the high energy and food prices during the period 2007–2011, also triggered high demand for land. The later unexpected fall in energy prices affected land use efficiency on land obtained by investors who typically took on large areas of land under long-term lease arrangements. Planting of energy crops is no longer profitable. Demand for land for food production by smallholder farmers is much less elastic.

Whether the development of a land sales market leads to a concentration of land in the hands of the few, and growing landlessness, is also an empirical question. Unequal land distributions are however more often the outcome of political processes, rather than redistributions through the land sales market. Some studies in Africa indicate that land purchases are made by land-poor but capital-rich persons and do not necessarily lead to more inegalitarian land distribution (Deininger and Mpuga 2008). This also implies that most owners of small farms are willing to sell their farms, and it may not be realistic in the foreseeable future to see a transition from small farms to medium-sized farms through an active land sales market in most African and many Asian countries. A recent study in Ethiopia (Holden and Bezu 2016) revealed a strong resistance against opening the land sales market and very high and increasing willingness to accept (WTA) prices for land.

One trigger towards the development of larger farm sizes appears to be rural wages. When the cost of labour in agriculture increases, we start to get a substitution into more capital-based production, as mechanisation becomes both labour- and cost-saving. There are limited or no economies of scale in most forms of tropical agriculture (Binswanger and Rosenzweig 1986; Binswanger et al. 1995), and hired labour is not a perfect substitute for family labour, thus giving family-based production a comparative advantage. Economies of scale exist in processing and marketing, and some forms of land development such as investment in irrigation, drainage and building of roads, and that is where large private companies can play an important role (Byerlee et al. 2015). However, even in land-abundant frontier areas, the most successful land development projects have involved subdivision of land for family farming (Byerlee et al. 2015).

In relation to the commercialisation of agriculture and development of the supply chain for agricultural products, such as the expansion of supermarkets, contract farming by smallholder producers has in many cases been quite successful. This is an area with a lot of scope for further development, and it could be an avenue to improving food security in a more urbanised world. In Asia, with economic development, a Westernisation of diets and a diversification away from strongly cereal-based diets towards much more varied diets can be observed, and obesity is taking over as the biggest nutrition-related problem (Pingali 2007; Popkin et al. 2012). The development of vertically integrated supply chains is part of this transformation, putting more pressure on the smallholder sector to commercialise, diversify and specialise production. There are more scale economies in some of these specialisations, such as poultry, and this can lead to larger farm sizes (Pingali 2007). More stringent quality standards can also push in the same direction, but this is queried and smallholders are found to supply a large share of the products in such supply chains (Reardon et al. 2009). Diversification of demand by wealthier consumers is also a key to high-value production and on-farm processing of such products. Farmers' markets in urban areas, supplying high-quality farm products, are also a way of raising returns on land and labour and intensifying production. This is a trend in developed countries, and it is likely to also expand in developing countries with the growth of a wealthy middle class.

Secure and well-defined property rights, and well-functioning land markets, are essential to create the incentives for investment in new types of production and obtain more optimally sized production units. While Africa is lagging behind Asia, it is likely to follow a similar evolutionary pathway of diversification and intensification in smallholder agriculture, as urbanisation expands and creates new market opportunities.

Land rental markets have been shown to be pro-poor in many contexts (Holden et al. 2008). Landless or near landless people may access land through renting if they have the necessary complementary resources to utilise the land efficiently. The land rental market can also transfer land from landlords who are poor in non-land resources, to tenants who are rich in non-land resources relative to land resources. Reverse tenancy systems, with poor landlords and relatively wealthier tenants, are dominant in Ethiopia, Eritrea, Madagascar and Tunisia. There has also been a general finding that land rental markets transfer land from less efficient to more efficient producers (Holden et al. 2008, 2013).

In areas where sharecropping is dominant and Marshallian inefficiency prevalent, the efficiency-enhancing effect of land renting is less clear. Otsuka (2007) associates Marshallian inefficiency with land-to-the-tiller reforms that have enhanced tenure insecurity of landowners in several Asian countries. Poor landlords have also been found to have weak bargaining power, and therefore to be less able to choose efficient tenants or enhance their production efficiency, in the reverse tenancy system in Ethiopia (Holden and Bezabih 2008; Ghebru and Holden 2015). Land rental contracts among kin partners have sometimes been associated with higher land use efficiency and in other cases with lower land use efficiency (Sadoulet et al. 1997; Kassie and Holden 2007; Ghebru and Holden 2015).

One may ask what the added value of land rental markets is in rural areas. The basic neoclassical farm household model (Singh et al. 1986) contained no land market and yielded no inefficiencies. It is when land is scarce and there are imperfections in non-land markets that re-allocation of resources through the land market can give efficiency gains, by reducing overall transaction costs and enhancing resource use efficiency.

How much agricultural land is reallocated through land rental markets varies substantially across countries. Land markets are likely to be non-existent or very thin in land-abundant countries. A skewed distribution of land relative to non-land resources, low elasticity of substitution among factors of production, and low transaction costs in land relative to non-land factor markets, will increase returns to participation in the land rental market to achieve balanced resource portfolios for production. An active land rental market is also likely to reduce incentives to sell land. It reduces the cost of ownership and enhances the flexibility, as management can be passed on to tenants while retaining land values that may be expected to increase over time with increasing land scarcity and economic development. It also reduces the need for capital in relation to accessing land. The rental market's advantage of reducing the need for credit for buying land may be more important than the usefulness of land as collateral in developing countries where the land sales market does not work well enough to favour the use of land as collateral. De Soto (2000) emphasised the need to formalise property rights in order to 'make dead capital alive', through linking the credit and land market by collateralising land. In developing agrarian economies, it is rather land rental markets that can reduce capital needs and enhance land use efficiency and growth in rural areas. The financial crisis also revealed that having too strong links between the credit and capital markets, through too much borrowing with security in immobile assets, can become the Achilles heel in the economy and contribute to economic instability. Rental markets for immobile assets can, on the other hand, help to create stability and reduce the need for distress sales of such assets.

Airbnb and other forms of short-term rentals are utilising modern information technologies (Internet) to dramatically reduce the transaction costs and information asymmetries in these markets. Transaction costs are still very high in land rental markets in developing countries, but there is scope for reducing these transaction costs and information asymmetries through better registries and information-sharing mechanisms that enhance transparency and accountability in tenure arrangements. This can reduce search, monitoring and enforcement costs for agents and government administrations.

6 Land Tenure Reforms, Tenure Security and Food Security

While classical land titling has given very limited benefits, or benefits very skewed in favour of the wealthy and well connected in many developing countries (Jacoby and Minten 2007; Benjaminsen et al. 2009), low-cost reforms to enhance tenure security in Ethiopia and Rwanda appear to have had significant positive effects, in terms of enhanced investments in soil conservation (both countries) and tree planting (Ethiopia), enhanced land productivity (Ethiopia) and land rental market activity (Ethiopia) (Deininger et al. 2008, 2011; Holden et al. 2009, 2011; Ali et al. 2014). In Ethiopia, there is also evidence of improved food security and child nutrition, particularly for female-headed households (Holden and Ghebru 2013; Ghebru and Holden 2013), while in Rwanda, land access for legally married women has improved and tenure security and land investments have increased for female-headed households (Ali et al. 2014). This is also likely to have had positive effects on food security in the case of Rwanda, although this has not been investigated by the researchers.

Maxwell and Wiebe (1999) and Holden and Ghebru (2016) have reviewed the literature relating land tenure and land tenure reforms to food security in developing countries. They demonstrate the complexity of such links, and that the literature on tenure issues and food security has largely involved separate fields of inquiry. Lawry et al. (2017) made a systematic review of studies of the impacts of land tenure reforms on investment and agricultural productivity; after reviewing 27,000 studies, they ended up with only 20 holding high-quality impact assessment standards. Ten of these were in Africa, five in Ethiopia. The external validity of these few high-quality studies may be questioned, given the diversity of tenure systems and their complex dynamics. There is clearly a need for more high-quality studies. But for policy purposes, we still need to draw on lower-quality studies. The urgency of issues limits how much we can rely on experimental designs that the 'randomistas' have argued to be the way forward in promoting economic development in developing countries. Such experiments are not a panacea solution in all-important developing country contexts where interventions are needed. Tenure systems are complex, and other forms of contextual and spatial heterogeneity are also important. We therefore need to pay careful attention to the underlying causal mechanisms, and we need structural models to capture more of the complexity, longer-run impacts and chain reactions related to tenure systems and tenure reforms (Deaton 2010).

There is clearly a need for more studies of the relationship between land tenure and food security in different contexts. The sharp increases in demand for land, related to the energy and food price boom in 2007–2010, revealed that investors obtained more land where tenure systems were poorly developed and population densities were low (Deininger and Byerlee 2012). However, there are also people with customary land rights in such areas, although their rights may not have been recognised by statutory law. Within the complex area of customary tenure, more attention needs to be paid to preventing the marginalisation of population groups with limited political influence, as their food and livelihood security can otherwise be threatened. While there is an urgent need to strengthen land governance in many countries, there is also a high risk that such reforms lead to elite capture given the trend in land distribution, converting customary land to state land for allocation to investors, as observed in a number of African countries (Jayne et al. 2016).

Important factors for whether customary land and smallholder farmers are protected from land acquisitions by investors are: (a) the extent to which customary land and the land rights of holders of customary land are recognised in statutory law;

(b) whether it is possible to convert customary land to state land and put it under long-term lease to investors; (c) whether customary land that has been converted and transferred to investors reverts to customary land at the end of the lease; (d) whether traditional leaders have the freedom to transfer customary land to investors and (e) whether the state can expropriate customary land for allocation to investors, possibly without any need to compensate those who have lost the land.

Alden Wily (2015) shows how a tenure reform in Côte d'Ivoire threatens to eliminate the land rights of people who have not been able to obtain a land title by a specific date. Other countries have attempted to integrate customary tenure systems into statutory law, e.g. by establishing customary tenure titles or certificates. Malawi is one of the countries now attempting this with its new Customary Land Bill, which was passed by Parliament in 2016. Other countries attempting formalisation of customary land rights in this way include Tanzania, Uganda and Zambia. Implementation has been slow in Uganda, while there is some anecdotal evidence of positive effects on tenure security and reduction of disputes in Zambia.

The transition from common property to individual land rights has been seen as a necessary part of the evolution of land rights with growing land scarcity, and there has been a tendency to associate customary tenure with common property. However, individual and household use rights to land can be quite strong in customary systems, although sales rights are often restricted. In studying land use in northern Zambia in low population density areas with long fallows and shifting cultivation in the 1980s, I was surprised myself that even the long fallow woodland areas had borders and belonged to individual households.

Alienation of the less influential also happens in customary systems, and women are particularly exposed in male-dominated customary systems (Peters 2004). Chiefs are usually in a dominant position in such systems, and lack of transparency and accountability has prompted demands for customary tenure reforms and establishment of more transparent and accountable land administrations and land conflict resolution systems. However, this could also be seen as a power struggle between a rural elite (traditional leaders) and an urban elite advocating for land tenure reform and being in a position to take advantage of such a reform. The extent to which the old system or the reform process will better protect the interests of the poor and vulnerable is an empirical issue. In this context, international organisations have an important role in safeguarding the interests of the poor, as well as in supporting reforms that contribute to economic and social development. Good systems for monitoring and impact assessment are essential and require specialised expertise that goes beyond the capacity of each country.

Ensuring food security and tenure security for rapidly growing urban populations is among the most demanding tasks ahead. Urban expansion is inevitable, and it may take place on high-potential agricultural land, leading to irreversible land losses for food production. Better long-term land use planning is crucial, to optimise the use of scarce land resources and to minimise production and transportation costs and greenhouse gas emissions. Good governance, with an emphasis on forward-looking legal and administrative systems, may easily be hijacked by short-term political interests and lead to conflicts that threaten social stability and welfare. 'Random walk' development of urban areas may have unforeseen future costs. This may be the case particularly in developing countries facing rapid population growth and urbanisation. Urban planning emphasising vertical rather than horizontal expansion can reduce the pressure on surrounding land areas. Providing young migrants with adequate housing, other social services and jobs is essential for inclusive development. Low-cost rental housing and labour-intensive construction businesses can help with the transition to urban environments for many young people. The Ethiopian youth group model is an interesting approach that may also be suitable in other countries and may work in both rural and urban settings (Holden and Tilahun 2017). It mobilises and organises youth, in rural as well as urban areas, in ways that enhance environmental conservation, resource utilisation, entrepreneurship, employment and infrastructure development.

7 Conclusions

Population growth, land degradation, climate change, urban expansion and changing diets are contributing to increasing land scarcity and strong demand for land use intensification to meet future food demands. Uncertainties in terms of timing and severity of agroclimatic and other shocks, including social and political instability, are threatening vulnerable developing economies and their systems of governance.

There is a need for a rapid transformation from rural to urban livelihoods for many in developing countries, while still facilitating absorption of more people on small farms in rural areas in the years to come. This is especially the case in Africa, where there are still large rural populations continuing to grow at a high rate. This will require formidable land use intensification through push and pull mechanisms. Tenure security, well-defined property rights and well-functioning property markets are essential for this transformation, in order to stimulate investment in land, diversification and intensification of production, provision of employment opportunities, affordable housing and other social services in rapidly growing urban areas.

Short-term as well as longer-term land rental and housing markets are essential to reduce the costs of transformation, to buffer shocks and facilitate social mobility. Good and competent governance, that facilitates the transformation through legal provision, and transparent and accountable land administration, will be crucial. National governance systems will need support from international organisations to build their competence and exchange ideas and experience, do careful research and develop better evidence-based policies.

New technologies for land registration, mapping, information dissemination and matching of suppliers and demanders in property markets have great potential, but there is also a high risk of misuse of such information by political elites and rent seekers. This is particularly the case in urban and peri-urban land markets, where property values are increasing very fast and there can be huge rents up for capture through legal as well as illegal processes. Small farms on the fringe of urban areas may see their land being expropriated while receiving minimal, if any, compensation.

The assignment of rights and the functioning of property markets and government regulation, including property taxation, in these urban and peri-urban areas will be immensely important to the future degree of inequity in wealth.

Small and even smaller farms will persist in rural areas in densely populated developing countries with high population growth. There is a high risk that rural poverty traps may grow in space and severity and trigger destitute migration in waves of growing scale. The migration flow from Africa and the Middle East into Europe over the last couple of years is only the beginning. Investment in smallholder agriculture and political stability in Africa could be one of the best ways to reduce such uncontrolled migration.

References

Alden Wily, L. (2015, March). Time to rethink? A critique of rural land law in Côte d'Ivoire. FERN Office UK and Brussels. www.fern.org.

Ali, D. A., & Deininger, K. (2015). Is there a farm size-productivity relationship in African agriculture? Evidence from Rwanda. *Land Economics, 91*(2), 317–343.

Ali, D. A., Deininger, K., & Goldstein, M. (2014). Environmental and gender impacts of land tenure regularization in Africa: Pilot evidence from Rwanda. *Journal of Development Economics, 110,* 262–275.

Ali, D. A., Deininger, K., & Harris, C. A. P. (2015). Using national statistics to increase transparency of large land acquisition: evidence from Ethiopia. World Bank policy Research Working Paper (p 7342).

Anseeuw, W., Jayne, T., Kachule, R., & Kotsopoulos, J. (2016). The quiet rise of medium-scale farms in Malawi. *Land, 5*(3), 19.

Anwar, T., Qureshi, S. K., Ali, H., & Ahmad, M. (2004). Landlessness and rural poverty in Pakistan [with comments]. *The Pakistan Development Review*, 855–874.

Balisacan, A. M. (1993). Agricultural growth, landlessness, off-farm employment, and rural poverty in the Philippines. *Economic Development and Cultural Change, 41*(3), 533–562.

Benjaminsen, T. A., Holden, S., Lund, C., & Sjaastad, E. (2009). Formalisation of land rights: Some empirical evidence from Mali, Niger and South Africa. *Land Use Policy, 26*(1), 28–35.

Bezu, S., & Holden, S. (2014). Are rural youth in Ethiopia abandoning agriculture? *World Development, 64,* 259–272.

Binswanger, H. P., Deininger, K., & Feder, G. (1995). Power, distortions, revolt and reform in agricultural land relations. *Handbook of Development Economics, 3,* 2659–2772.

Binswanger, H. P., & Rosenzweig, M. R. (1986). Behavioural and material determinants of production relations in agriculture. *The Journal of Development Studies, 22*(3), 503–539.

Byerlee, D., Masters, W. A., & Robinson, D. (2015). From land grab to land development: The past and potential of private investment in frontier agriculture. Paper presented at the World Bank Land and Poverty conference, March 23–27, 2015, Washington, DC.

Carletto, C., Savastano, S., & Zezza, A. (2013). Fact or artifact: the impact of measurement errors on the farm size—productivity relationship. *Journal of Development Economics, 103,* 254–261.

de Schutter, O. (2011). How not to think of land-grabbing: Three critiques of large-scale investments in farmland. *The Journal of Peasant Studies, 38*(2), 249–279.

de Soto, H. (2000). *The mystery of capital: Why capitalism triumphs in the West and fails everywhere else.* Basic books.

Deaton, A. (2010). Instruments, randomization, and learning about development. *Journal of Economic Literature, 48*(2), 424–455.

Deininger, K., Ali, D. A., & Alemu, T. (2011). Impacts of land certification on tenure security, investment, and land market participation: evidence from Ethiopia. *Land Economics, 87*(2), 312–334.

Deininger, K., Ali, D. A., Holden, S., & Zevenbergen, J. (2008). Rural land certification in Ethiopia: Process, initial impact, and implications for other African countries. *World Development, 36*(10), 1786–1812.

Deininger, K., & Byerlee, D. (2012). The rise of large farms in land abundant countries: Do they have a future? *World Development, 40*(4), 701–714.

Deininger, K., Hilhorst, T., & Songwe, V. (2014). Identifying and addressing land governance constraints to support intensification and land market operation: Evidence from 10 African countries. *Food Policy, 48,* 76–87.

Deininger, K., & Mpuga, P. (2008). Land markets in Uganda: What is their impact and who benefits? In S. T. Holden, K. Otsuka, & F. M. Place (Eds.), *The emergence of land markets in Africa: Impacts on poverty, equity, and efficiency* (pp. 131–155). Washington DC: RFF Press/Routledge.

German, L., Schoneveld, G., & Mwangi, E. (2013). Contemporary processes of large-scale land acquisition in Sub-Saharan Africa: Legal deficiency or elite capture of the rule of law? *World Development, 48,* 1–18.

Ghebru, H., & Holden, S. T. (2013). Links between tenure security and food security: Evidence from Ethiopia. CLTS Working Paper No. 2/2013. Centre for Land Tenure Studies, Norwegian University of Life Sciences, Ås, Norway.

Ghebru, H. H., & Holden, S. T. (2015). Reverse-share-tenancy and agricultural efficiency: Farm-level evidence from Ethiopia. *Journal of African Economies, 24*(1), 148–171.

HLPE. (2013). Investing in smallholder agriculture for food security. A Report by the High Level Panel of Experts on Food Security and Nutrition (Vol. 6). Rome: FAO.

Holden, S. T., Bezabih, M. (2008). Gender and land productivity on rented land in Ethiopia. In: S. T. Holden, K. Otsuka, F. Place (Eds.), *The emergence of land markets in Africa: Assessing the impacts on poverty, equity and efficiency* (pp. 179–98).

Holden, S. T., & Bezu, S. (2016). Preferences for land sales legalization and land values in Ethiopia. *Land Use Policy, 52,* 410–421.

Holden, S. T., & Binswanger, H. P. (1998). Small farmer decision-making, market imperfections, and natural resource management in developing countries. In: *Agriculture and the environment: Perspectives on sustainable rural development* (pp. 50–71). Washington, DC: The World Bank.

Holden, S. T., Deininger, K., & Ghebru, H. (2009). Impacts of low-cost land certification on investment and productivity. *American Journal of Agricultural Economics, 91*(2), 359–373.

Holden, S. T., Deininger, K., & Ghebru, H. (2011). Tenure insecurity, gender, low-cost land certification and land rental market participation. *Journal of Development Studies, 47*(1), 31–47.

Holden, S. T., & Ghebru, H. (2013). Welfare impacts of land certification in Tigray, Ethiopia. In S. T. Holden, K. Otsuka, & K. Deininger (Eds.), *Land tenure reform in Asia and Africa: Assessing impacts on poverty and natural resource management* (pp. 137–161). London: Palgrave Macmillan.

Holden, S. T., & Ghebru, H. (2016). Land tenure reforms, tenure security and food security in poor agrarian economies: Causal linkages and research gaps. *Global Food Security, 10,* 21–28.

Holden, S. T., Otsuka, K., & Deininger, K. (Eds.). (2013). *Land tenure reform in Asia and Africa: Assessing impacts on poverty and natural resource management.* London: Palgrave Macmillan.

Holden, S. T., Otsuka, K., & Place, F. M. (Eds.). (2008). *The emergence of land markets in Africa: impacts on poverty, equity, and efficiency*. Washington, DC: RFF Press/Routledge.

Holden, S., & Shiferaw, B. (2004). Land degradation, drought and food security in a less-favoured area in the Ethiopian highlands: A bio-economic model with market imperfections. *Agricultural Economics, 30*(1), 31–49.

Holden, S. T., Shiferaw, B., & Wik, M. (1998). Poverty, market imperfections, and time preferences: Of relevance for environmental policy? *Environment and Development Economics, 3,* 105–130.

Holden, S. T., & Tilahun, M. (2018). The importance of Ostrom's design principles: Youth group performance in Northern Ethiopia. *World Development, 104*, 10–30.

Jacoby, H. G., & Minten, B. (2007). Is land titling in Sub-Saharan Africa cost-effective? Evidence from Madagascar. *The World Bank Economic Review, 21*(3), 461–485.

Jayne, T. S., Chamberlin, J., & Headey, D. D. (2014). Land pressures, the evolution of farming systems, and development strategies in Africa: A synthesis. *Food Policy, 48,* 1–17.

Jayne, T. S., Chamberlin, J., Traub, L., Sitko, N., Muyanga, M., Yeboah, F. K., et al. (2016). Africa's changing farm size distribution patterns: the rise of medium-scale farms. *Agricultural Economics, 47*(S1), 197–214.

Kassie, M., & Holden, S. T. (2007). Sharecropping efficiency in Ethiopia: Threats of eviction and kinship. *Agricultural Economics, 37*(2–3), 179–188.

Lanjouw, P. (2007). Does the rural nonfarm economy contribute to poverty reduction. In S. Haggblade, P. B. Hazell, & T. Reardon (Eds.), *Transforming the rural nonfarm economy: Opportunities and threats in the developing world* (pp. 55–79). Washington, DC: International Food Policy Research Institute.

Lawry, S., Samii, C., Hall, R., Leopold, A., Hornby, D., & Mtero, F. (2017). The impact of land property rights interventions on investment and agricultural productivity in developing countries: A systematic review. *Journal of Development Effectiveness, 9*(1), 61–81.

Masters, W. A., Djurfeldt, A. A., De Haan, C., Hazell, P., Jayne, T., & Jirström, M. (2013). Urbanization and farm size in Asia and Africa: Implications for food security and agricultural research. *Global Food Security, 2*(3), 156–165.

Maxwell, D., & Wiebe, K. (1999). Land tenure and food security: Exploring dynamic linkages. *Development and Change, 30*(4), 825–849.

Meenakshi, J. V., & Ray, R. (2002). Impact of household size and family composition on poverty in rural India. *Journal of Policy Modeling, 24*(6), 539–559.

Otsuka, K. (2007). Efficiency and equity effects of land markets. *Handbook of Agricultural Economics, 3,* 2671–2703.

Peters, P. E. (2004). Inequality and social conflict over land in Africa. *Journal of Agrarian Change, 4*(3), 269–314.

Pingali, P. (2007). Westernization of Asian diets and the transformation of food systems: Implications for research and policy. *Food Policy, 32*(3), 281–298.

Popkin, B. M., Adair, L. S., & Ng, S. W. (2012). Global nutrition transition and the pandemic of obesity in developing countries. *Nutrition Reviews, 70*(1), 3–21.

Ravallion, M., & Van de Walle, D. (2008). Does rising landlessness signal success or failure for Vietnam's agrarian transition? *Journal of Development Economics, 87*(2), 191–209.

Reardon, T., Barrett, C. B., Berdegue, J. A., & Swinnen, J. (2009). Agrifood industry transformation and farmers in developing countries. *World Development, 37*(11), 1717–1727.

Sadoulet, E., De Janvry, A., & Fukui, S. (1997). The meaning of kinship in sharecropping contracts. *American Journal of Agricultural Economics, 79*(2), 394–406.

Shiferaw, B., & Holden, S. T. (1998). Resource degradation and adoption of land conservation technologies in the Ethiopian highlands: A case study in Andit Tid, North Shewa. *Agricultural Economics, 18*(3), 233–247.

Simtowe, F., Mendola, M., Mangisoni, J., Tchale, H., & Nyirongo, C. (2013). The economic effects of land redistribution: The case of a community-based rural land development project in Malawi. In S. T. Holden, K. Otsuka, & K. Deininger (Eds.), *Land tenure reform in Asia and Africa: Assessing impacts on poverty and natural resource management* (pp. 105–133). London: Palgrave Macmillan.

Singh, I., Squire, L., & Strauss, J. (1986). *Agricultural household models: Extensions, applications, and policy*. Washington, DC: The World Bank.

Sitko, N. J., & Jayne, T. S. (2014). Structural transformation or elite land capture? The growth of "emergent" farmers in Zambia. *Food Policy, 48,* 194–202.

United Nations. (2014). *World urbanization prospects: The 2013 revision*. New York: United Nations Department of Economic and Social Affairs.

PERMISSIONS

All chapters in this book were first published by Springer; hereby published with permission under the Creative Commons Attribution License or equivalent. Every chapter published in this book has been scrutinized by our experts. Their significance has been extensively debated. The topics covered herein carry significant findings which will fuel the growth of the discipline. They may even be implemented as practical applications or may be referred to as a beginning point for another development.

The contributors of this book come from diverse backgrounds, making this book a truly international effort. This book will bring forth new frontiers with its revolutionizing research information and detailed analysis of the nascent developments around the world.

We would like to thank all the contributing authors for lending their expertise to make the book truly unique. They have played a crucial role in the development of this book. Without their invaluable contributions this book wouldn't have been possible. They have made vital efforts to compile up to date information on the varied aspects of this subject to make this book a valuable addition to the collection of many professionals and students.

This book was conceptualized with the vision of imparting up-to-date information and advanced data in this field. To ensure the same, a matchless editorial board was set up. Every individual on the board went through rigorous rounds of assessment to prove their worth. After which they invested a large part of their time researching and compiling the most relevant data for our readers.

The editorial board has been involved in producing this book since its inception. They have spent rigorous hours researching and exploring the diverse topics which have resulted in the successful publishing of this book. They have passed on their knowledge of decades through this book. To expedite this challenging task, the publisher supported the team at every step. A small team of assistant editors was also appointed to further simplify the editing procedure and attain best results for the readers.

Apart from the editorial board, the designing team has also invested a significant amount of their time in understanding the subject and creating the most relevant covers. They scrutinized every image to scout for the most suitable representation of the subject and create an appropriate cover for the book.

The publishing team has been an ardent support to the editorial, designing and production team. Their endless efforts to recruit the best for this project, has resulted in the accomplishment of this book. They are a veteran in the field of academics and their pool of knowledge is as vast as their experience in printing. Their expertise and guidance has proved useful at every step. Their uncompromising quality standards have made this book an exceptional effort. Their encouragement from time to time has been an inspiration for everyone.

The publisher and the editorial board hope that this book will prove to be a valuable piece of knowledge for researchers, students, practitioners and scholars across the globe.

List of Contributors

Jacob Ricker-Gilbert
Department of Agricultural Economics, Purdue University, West Lafayette, IN, USA

Augustine Langyintuo
International Finance Corporation, World Bank Group, Delta Center, Menengai Road , Upper Hill, Nairobi, Kenya

Alwin Dsouza and Ashok K. Mishra
Arizona State University, Mesa, AZ, USA

Munir A. Hanjra
International Water Management Institute (IWMI), Pretoria, South Africa

Timothy O. Williams
International Water Management Institute (IWMI), Accra, Ghana

Shenggen Fan
International Food Policy Institute (IFPRI), Washington, DC, USA
China Agricultural University, College of Economics and Management, Beijing, China

Christopher Rue
International Food Policy Institute (IFPRI), Washington, DC, USA

Kelsey Shaw
Coherent Economics, Chicago 60602, USA

Sébastien Mary
Department of Economics, Driehaus College of Business, DePaul University, Chicago, USA

Peter Hazell
Santa Barbara, CA, USA

Mathew Abraham and Prabhu Pingali
Tata-Cornell Institute for Agriculture and Nutrition, The Charles H. Dyson School of Applied Economics and Management, Cornell University, Ithaca, New York, USA

Donald F. Larson
Global Research Institution, College of William and Mary, Williamsburg, USA

Rie Muraoka
Graduate Institute for Policy Studies, Japan Society for the Promotion of Science, Tokyo, Japan

Tadashi Sonoda
Nagoya University, Nagoya, Japan

Keijiro Otsuka
Graduate School of Economics, Kobe University, Kobe, Japan

Stein T. Holden
School of Economics and Business, Centre for Land Tenure Studies, Norwegian University of Life Sciences, 1432 Ås, Norway

Index

Printed in the USA
CPSIA information can be obtained
at www.ICGtesting.com
LVHW010815221023
761596LV00005B/512